# Exploitation of A Ship's Magnetic Field Signatures

Exploitation of A Ship's Magnetic Field Signatures

John J. Holmes

978-3-031-00565-7    paperback
978-3-031-00565-7    paper

978-3-031-01693-6    ebook
978-3-031-01693-6    ebook

DOI    10.1007/978-3-031-01693-6

A Publication in the Springer series
*SYNTHESIS LECTURES ON COMPUTATIONAL ELECTROMAGNETICS #9*
Series Editors: Constantine A, Balanis, Arizona State University

1932-1252    Print
1932-1716    Electronic

First Edition

10 9 8 7 6 5 4 3 2 1

# Exploitation of A Ship's Magnetic Field Signatures

**John J. Holmes**
Naval Surface Warfare Center,
West Bethesda, Maryland, USA

*SYNTHESIS LECTURES ON COMPUTATIONAL ELECTROMAGNETICS #9*

*To Marlene and Bud*

# ABSTRACT

Surface ship and submarine magnetic field signatures have been exploited for over 80 years by naval influence mines, and both underwater and airborne surveillance systems. The generating mechanism of the four major shipboard sources of magnetic fields is explained, along with a detailed description of the induced and permanent ferromagnetic signature characteristics. A brief historical summary of magnetic naval mine development during World War II is followed by a discussion of important improvements found in modern weapons, including an explanation of the damage mechanism for non-contact explosions. A strategy for selecting an optimum mine actuation threshold is given. A multi-layered defensive strategy against naval mines is outlined, with graphical explanations of the relationships between ship signature reduction and minefield clearing effectiveness. In addition to a brief historical discussion of underwater and airborne submarine surveillance systems and magnetic field sensing principles, mathematical formulations are presented for computing the expected target signal strengths and noise levels for several barrier types. Besides the sensor self-noise, equations for estimating geomagnetic, ocean surface wave, platform, and vector sensor motion noises will be given along with simple algorithms for their reduction.

# KEYWORDS

Magnetic anomaly detection, Magnetic signatures, Naval mines, Submarine surveillance

# Permissions

All graphics and photographs in this manuscript were produced by the author, or obtained from various internet web sites.

# Contents

1. Introduction ........................................................... 1

2. Shipboard Sources of Magnetic Field ................................. 5
   2.1 Evolution of the Magnetic Ship ................................. 5
   2.2 Ferromagnetic Signatures ...................................... 6
   2.3 Other Important Sources of Magnetic Field .................... 14

3. Exploitation of Magnetic Signatures by Naval Mines ................ 19
   3.1 Evolution of the Magnetic Mine ............................... 19
   3.2 Modern Magnetic Mine Technologies ........................... 21
   3.3 Magnetic Mine Countermeasures ............................... 29

4. Exploitation of Magnetic Signatures by Submarine Surveillance Systems ........ 39
   4.1 Evolution of the Submarine Magnetic Detection System ........ 39
   4.2 Magnetic Harbor Loops ....................................... 41
   4.3 Submarine Barriers Using Triaxial Magnetic Field Vector Sensors ........... 48
   4.4 Submarine Barriers Using Total Field Magnetometers .......... 52

5. Summary ............................................................. 61

# Acknowledgments

The author would like to sincerely thank Mr. John Scarzello for reviewing this manuscript, and for his discussions on underwater electromagnetic sensor technologies. He also appreciates the signature processing efforts of Mr. Daniel Lenko, and for the comments provided by Mr. Walter Dence.

# CHAPTER 1

# Introduction

Detecting the presence of a naval vessel by sensing its underwater magnetic field has been exploited primarily in the arena of *undersea warfare*. Owing to propagation losses of electromagnetic fields in electrically conducting seawater the frequencies of interest are generally limited to those in the ultra-low frequency (ULF) spectrum of ~0 to 3 Hz, and the extremely low frequency (ELF) band from 3 Hz to 3 kHz. Until recently, military systems used in undersea warfare focused almost entirely on the ULF and low end of the ELF bands. Although the decision to build an "all electric" ship that uses high power electric propulsion motors and generators, and high current distribution systems has increased the importance of the ELF band, this discussion will focus primarily on exploitation of a surface ship's or submarine's ULF magnetic field signatures.

The term "*signature*" was coined in the realm of acoustic measurement and detection of a vessel's underwater sound pressure field. Like handwriting on a signed document, underwater sound produced by a ship has characteristics that are unique to it, and can be used to distinguish it from others. Even though a surface ship's or submarine's magnetic field is not as unique as its acoustic, the term "signature" has been carried over to describe the spatial and temporal distribution of a ship's electromagnetic fields.

Most people are familiar with the patterns made by slivers or filings of iron sprinkled around a magnet. When spread over a cardboard sheet placed on top of a magnet and vibrated, the magnetized iron filings align with the magnetic field. The resulting patterns trace the contours of the magnetic field that can not be seen directly. Because this magnetic field is invisible to the eye and can not be heard or felt, it is difficult, at times, to understand how its presence around naval vessels increases the danger of attack by mines and detection systems. A rudimentary understanding of the shipboard generating mechanisms of magnetic field signatures, and the physics behind sensing them should reduce some of the mystery.

There are four primary ship borne sources of magnetic field in the ULF band. They are:

1. Ferromagnetism induced by the earth's natural magnetic field in the ferrous steel used to construct naval vessels.

2. Eddy currents induced in any shipboard electrically conducting material (magnetic as well as non-magnetic) as it rotates in the earth's magnetic field.

3. Electric currents impressed into a ship's conducting hull and the surrounding seawater by natural electrochemical corrosion processes or by cathodic protection systems designed to prevent the ship from corroding (rusting).

4. Currents that flow in electric motors, generators, distribution cables, switch gear, breakers, and other active circuits found onboard.

The physical processes behind each ship borne source of magnetic field will be described in more detail in Chapter 2, along with a quick outline of methods to reduce them.

Exploitation of a ship's or submarine's magnetic field signatures in undersea warfare falls into two subcategories; *mine warfare* (MIW) and *anti-submarine warfare* (ASW). Mines were the first weapons designed to detect, locate, and attack a vessel by sensing its magnetic field. In 1920, Germany started the large-scale development of a sea mine that rests on the ocean floor or floats in the water column until it detects the magnetic field of a target ship and then, if certain requirements have been met, detonates. It was discovered that the shock wave produced by a non-contact underwater explosion was capable of sinking or heavily damaging a ship at a distance. These non-contact weapons are called *magnetic influence mines*, and their basic operating principles will be covered in Chapter 3.

Mines are dangerous. Naval mines have sunk or damaged ships *whose combined weight exceeds* hundreds of thousands of tonnes. During World War I, a total of 309 700 mines were laid by all sides, sinking or damaging more than 950 vessels. These figures increased to 700 000 mines during World War II, placing the ship lost or damaged figure to over 3200 [1]. Since 1950, 14 U.S. naval vessels have been casualties of the mine threat, the most recent being the *USS PRINCETON* (CG 59) that was damaged in 1991 by an Iraqi magnetic mine during Operation Desert Storm [2]. Even as recently as Operation Iraqi Freedom (2003), coalition navies were faced with the possibility of a significant mine-clearing operation, which was circumvented when most of the weapons were captured before they could be deployed by Iraqi forces [3].

Mines are cheap and can be easily manufactured or bought in the international weapons market. They are difficult and time consuming to find and neutralize, and can be deployed by an adversary using covert means without directly confronting the strength of a heavily armed naval vessel. Suffering causalities from a naval minefield could cost the lives of sailors, delay or alter the outcome of a conflict, prevent rapid reconstitution of naval capabilities due to minefield causalities, hurt economies, and adversely influence foreign and domestic politics. Unabated, naval mines are a very effective weapon, and as such, action must be taken to counter their effectiveness.

Several tools are available today to defeat the moored and bottom magnetic influence mine. *Mine sweeping*, *mine hunting*, and *ship signature reduction* are used in combination to reduce the probability of a combatant or support ship actuating an influence mine. These *mine countermeasure* (MCM) technologies will be covered in Chapter 3 to the extent of showing how they complement each other and produce synergistic benefits. Reducing the susceptibility of naval vessels to actuating mines while minimizing the timeline and material resources needed to render a minefield ineffective is the focus of the MCM research.

It is neither the objective, nor is it possible here, to give a comprehensive discussion on all aspects of offensive and defensive mine warfare. There are many factors that must be considered in developing an offensive influence mine, and even more when countering such a threat, whose order of importance depends heavily on the scenario. Instead, the top level technical aspects of designing and using magnetic influence mines will be described, along with systems and techniques to protect ones own fleet from these weapons.

A second important category of undersea warfare, where a vessel's magnetic field signature can be exploited, is in ASW. In the past, active and passive sonar systems were the primary means to detect submerged submarines. However, technologies to quieten the sound emanating from a submarine have been proliferated, while ASW scenarios of concern have shifted to the high-noise and acoustically challenging shallow water littoral ocean environments. As a result, ASW missions have changed to a point where the detection range of a submarine's magnetic field is now competitive with acoustic techniques. Technologies to detect and locate a submerged submarine by its magnetic field signature will be discussed in Chapter 4.

## REFERENCES

[1]  G. K. Hartmann and S. C. Truver, *Weapons That Wait*, 2nd ed, Annapolis, MD: Naval Institute Press, 1991, pp. 242–244.

[2]  Chief of Naval Operations. "Thunder and Lightning: The War with Iraq," Department of the Navy-Naval Historical Center, Washington, DC, May, 1991 [Online]. Available: http://www.history.navy.mil/wars/dstorm/ds5.htm.

[3]  P. J. Ryan, "Iraqi freedom: Mine countermeasures a success," *Proc. U.S. Naval Inst.*, vol. 129/5/1,203, May 2003.

CHAPTER 2

# Shipboard Sources of Magnetic Field

## 2.1 EVOLUTION OF THE MAGNETIC SHIP

Ships have not always been made of steel. Up until the middle of the 19th Century both commercial vessels and military warships were constructed entirely of wood, as they had been throughout history. However, the adaptation of cannon artillery to shipboard use had become very effective during close-in bombardment of both shore installations and other vessels. A large artillery shell could easily rip open the wooden hull of a vessel inflicting heavy damage to its structure, and sometimes even sinking it. As iron plating became more readly available its naval application as a protective armor was inevitable.

The first ship to cover its wooden hull with iron plates was the French warship *La Gloire*. *La Gloire* weighed 5630 tons and was known as an *ironclad* since her iron armor was attached or cladded to the outside of her underlying wooden hull [1]. She was commissioned in 1859, and was followed a few months later in 1860 by *HMS WARRIOR*, the first combatant with a completely iron hull. Commissioned in 1860 by the British Royal Naval, the hull of *WARRIOR* was constructed entirely of iron instead of cladding over wood, and weighed 9210 tons [2]. Although these two ships were the first to be protected with iron, it wasn't until the American Civil War that the military worth of naval armor was clearly demonstrated.

The first naval artillery exchange between two ironclad vessels occurred in 1862 during the American Civil War at the *Battle of Hampton Roads*. The 3200 ton confederate ironclad, *CSS VIRGINIA* [3], which was also called the *MERRIMAC*, exchanged gun fire with the U.S. navy's 980 ton armor-protected warship, *USS MONITOR* [4]. Although the two combatants lobbed shells at each other for hours, neither ship could be declared victorious. In spite of this, the battle of the *MONITOR* and *MERRIMAC* did conclusively demonstrate the superiority of armored warships over those with wooden hulls.

By the turn of the century nearly all frontline naval combatants were constructed from steel. These early battleships and battle-cruisers were called *dreadnoughts*, after the 18 000 ton technically revolutionary warship of its time, *HMS DREADNOUGHT*. To protect these mammoth ships from explosive artillery shells their hulls were built of ferrous steel up to 3 inches thick.

As is often the case, the countermeasure to one military threat results in a vulnerability to another. During World War I, the British Royal Navy attempted to exploit the ferromagnetic properties of a combatant's steel hull and introduced the first naval mine that triggered off a ship's magnetic field. Although the finicky mechanical nature of this mine's firing mechanism prevented it from having an impact on the war, the magnetic mine-genie was released from its bottle and set loose upon naval warfare.

Today, nearly all naval combatants are built from ferrous steel. The permeability constants of ferromagnetic steels used in modern naval construction can approach 300, which allow their thick hull to be low reluctance paths for the earth's static magnetic field, distorting it in the process. This distortion or anomaly in the earth's field can be detected as a time varying signal when the vessel sails over a stationary sensor, or one mounted on a moving platform. This monograph will provide an overview of shipboard sources of magnetic field signatures, and their exploitation by naval mines and submarine surveillance systems.

## 2.2  FERROMAGNETIC SIGNATURES

The first and most important shipboard source of magnetic field is magnetization of the ferromagnetic steel used in the construction of the hull, internal structure, machinery, and equipment items of the naval vessel. The earth's natural magnetic field can be easily detected with a pocket compass, and is responsible for inducing the magnetization in a ship. Although the highly non-linear magnetohydrodynamics that takes place in the earth's core is not completely understood [5], the distribution of its main magnetic field over the world's oceans has been mapped extensively and modeled [6], and is used in the prediction of a vessel's magnetic signature as a function of latitude, longitude, and heading.

Magnetic fields are produced by the electric charges in motion. The charges can be flowing in a linear fashion as an electric current, or they can be spinning about their own axis. In addition, electric charges are either positive, as in the case of a proton, or negative like an electron. The primary source of all ship and submarine magnetic signatures is the movement of the negatively charged electrons.

The explanation of the microscopic generating mechanism of a magnetic field by a moving electric charge is rooted in the theory of relativity [7]. In addition, the quantum mechanical explanation of the electron charge is still not fully understood. Nevertheless, a macroscopic model of a flowing or spinning electron as a uniformly charged sphere is sufficient to explain the magnetic fields from ships.

When an electric current flows in a wire magnetic fields circulate around it. As electrons flow through a conductor from the negative side of a battery to the positive they leave behind positively charged atoms, which are fixed in the crystalline structure of the metal conductor and do not move. The positively charged atoms vacated by electrons are called *holes*, and can be treated as positive charges moving in the direction opposite to the electron flow. The direction of positive charge flow is called *conventional current* and travels from the positive battery pole, through the circuit to the negative pole. The magnetic field generated by the electric current circulates around the wire according to the right-hand rule. (Place the thumb of the right hand in the direction of the conventional current flow, and then curling the fingers gives the direction of magnetic field.) A magnetic field exists at all points along the circuit regardless of its geometry.

The magnetic field of a current loop begins to take on a regular shape as it shrinks in size. In the limit as the single loop shrinks toward a point its magnetic field is predicted by the equations of a dipole with a moment $\bar{m}$ equal to the loop area times its current [8]. If the loop contains multiple wires the total dipole moment is the sum of those computed for each conductor. The unit of magnetic dipole moment in the *SI* system is ampere-meter$^2$, while that of the field intensity $\bar{H}$ is amperes/meter and the flux density $\bar{B}$ is tesla.

A spinning electron also produces a dipolar magnetic field and is the underlying source of ferromagnetism. Since an electron is negatively charged, its magnetic field appears to be generated by a loop with conventional (positive) current flowing in the direction opposite to its rotation. The magnetic dipole moment of a single spinning electron is approximately $9.27 \times 10^{-24}$ A-m$^2$. Electrons in the orbits around the nucleus of an atom can rotate about their own axis in either of the two directions, called up or down spin. In the atoms of ferromagnetic elements such as iron, all orbital shells, except the 3d, are filled with equal numbers of up and down spinning electrons. The unpaired electrons in the $3d$ orbit result in a net non–zero magnetic spin moment, whose fields can influence unpaired $3d$ electrons in adjacent atoms of the crystal.

An element having unpaired electrons in its $3d$ orbit is not sufficient to make it ferromagnetic. The distance between neighboring atoms in a crystalline structure must be right for the favorable exchange of energy between these unpaired electrons to affect each other's spin. Those elements with a positive energy exchange tend to be ferromagnetic including iron, the major element in ship steels.

Alloying of elements can change their ferromagnetic properties by changing their crystalline spacing. For example, if manganese is alloyed with copper, aluminum and tin its atomic spacing is increased, resulting in ferromagnetism even though none of the elements are by themselves ferromagnetic. Conversely, if iron is alloyed with chromium and nickel the resulting steel can be made non-magnetic since its atomic spacing does not support a favorable exchange of energy between atoms. As would be expected, the use of aluminum and non-magnetic steels in the construction of naval vessels to reduce their magnetic signatures is an important application.

Steels with high amounts of chromium are called stainless steels. Various types of stainless steels can be produced by changing the proportions of iron, chromium, nickel, carbon, and other elements in the alloying process. Not all stainless steels, however, are non-magnetic.

Some *martensitic* stainless steels (higher carbon stainless steels) are still very ferromagnetic, while some *austentic* stainless steels (higher chromium content) can have a very low magnetic permeability. However, the cost of austentic stainless steel, their requirement for special welding procedures, and corrosion issues prevent them from becoming an immediate replacement for ferromagnetic naval steels.

A piece of unmagnetized ferromagnetic material can have small sections where unpaired electron dipoles of neighboring atoms are all aligned in the same direction. This forms what is called a *magnetic domain*. For example, inside iron or steel armor plate many of these microscopic magnetic domains exist, but they point in different random directions when the sample is unmagnetized or *demagnetized*. When a magnetic field is imposed on the sample some of its domains become aligned in the direction of the applied field, while others remain in their original orientation resulting in a partially magnetized sample. If the applied field is strong enough, then all the ferromagnetic domains and unpaired spinning electrons align themselves in the same direction as that of the magnetic field. This fully-magnetized condition is called *magnetic saturation* and can be thought of as one large magnetic domain covering the entire sample. Saturation is the upper limit on magnetizing any ferromagnetic material since all unpaired electrons in the $3d$ orbits of all its atoms are now aligned.

Magnetizing and demagnetizing ferromagnetic steel is a very non-linear and complicated process. Once again a simple model will be used here to explain the magnetization process. Details of the physics surrounding ferromagnetism can be found in the classic book by Bozorth [9].

Unmagnetized steel, with its randomized domain orientations, will remain in this state only if it is shielded from all sources of magnetic field. With no external field applied to demagnetized steel, the net internal flux density and magnetization are zero, and is labeled as point 1 at the origin of the curve in Fig. 2.1. Applying a small positive external field to the steel, its internal flux density and magnetization increase (point 2 of Fig. 2.1), with some of its magnetic domains aligning in the direction of the applied field. When the external field has been reduced back to zero the domains return to their original zero magnetization orientation. This reversible magnetization is proportional to the external applied field and is called *induced magnetization*. The proportionality constant between the internal *B-field* and the external applied *H-field* is the slope of the line from point 1 to 2, and is called the *magnetic permeability*. The permeability has units of henrys/meter, and is a characteristic of the specific alloy of steel.

The steel's magnetization will not return to zero if the applied field is strong enough. For example, if the external field is increased sufficiently to reach point 3 in Fig. 2.1 and then removed, the magnetization will not return to zero. Instead, some of the magnetic domains aligned with the larger applied field will remain in that orientation after the inducing field is removed. The magnetization follows a curve that takes it to point 4 on the ordinate instead of back to the origin. The magnetization that remains in the steel when the inducing field has been

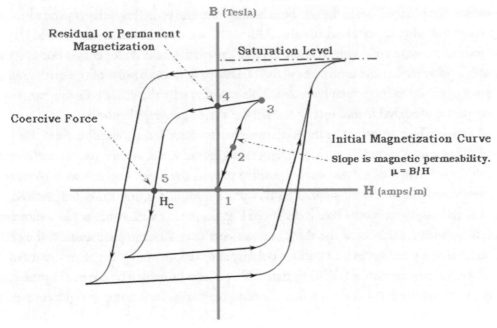

**FIGURE 2.1:** Example hysteresis curve for ferromagnetic material

removed is called the residual or *permanent magnetization*. To force the steel's magnetization back to zero a negative magnetic field must be applied of sufficient strength to reach point 5. The magnitude of applied field at this point is called the *coercive force*, and is also a characteristic of the steel alloy. It should be noted that although the magnetization of the steel is zero, it is not demagnetized since a field of $H_C$ must be continuously applied to maintain this state. Cycling the applied field through increasingly larger positive and negative levels produces a continuous family of *hysteresis* curves, of which only a few are drawn in the figure. The positive saturation level is indicated on the example hysteresis curve of Fig 2.1, with a corresponding negative saturation level in the third quadrant.

When a steel bar is placed in a uniform field not only is it magnetized as explained, but this magnetization also distorts the inducing field causing its flux lines to bend toward the steel. By convention, the magnetic field leaves the north pole of a magnet and enters its south pole. If the magnetized steel is moved across a sensor or if a sensor is moved past the steel, the anomalous field can be measured as a time-varying signal. This occurs whether the field is produced by an induced magnetization, permanent magnetization, or a combination of the two. It is this time-varying field that is exploited by magnetic mines and submarine surveillance systems.

The magnitude of the earth's magnetic field is on the order of 50 000 nT, and may be taken as uniform over the length of any ship. The earth's field can be separated into a vertical component that points down in the northern hemisphere and up in the southern, and a horizontal component that always points northward. Although by convention a magnetic field leaves the north pole of a magnet and enters its south pole, the earth's field vector comes out of its south geographic polar region and enters its north. The north and south poles of the earth's magnetic dipole are opposite to its geographic poles. This explains why the north pole of a magnet used as a compass is attracted to and rotates toward the earths geographic north.

A ship can be magnetized by the earth in each of its three orthogonal directions. Each magnetization state in-turn produces three magnetic signature vectors called the *vertical component* (positive down), *longitudinal component* (positive toward the bow), and *athwartship component* (positive toward the starboard side). The flux pattern around a uniformly magnetized vessel located at the magnetic North Pole is drawn in Fig. 2.2, along with contour plots showing the complete signature patterns of the three components over a rectangular area on the seafloor. Light areas in the plots represent a positive polarity and dark areas negative. Familiarity with the *induced vertical magnetization* (IVM) signature shapes can be gained by comparing the contour patterns to those expected from the flux distribution drawn in the upper right corner of the figure.

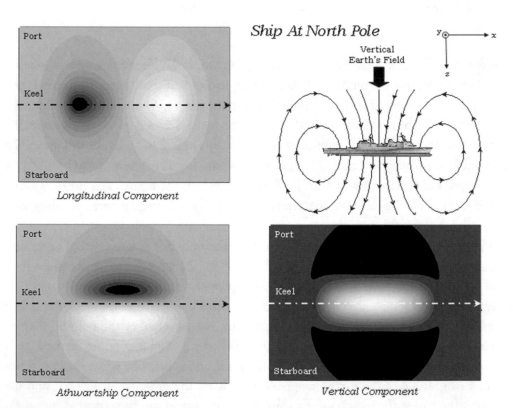

**FIGURE 2.2:** Induced magnetic field signature components of a vertically magnetized ship

When a ship is sailing north at the magnetic equator it receives an *induced longitudinal magnetization* (ILM) from the earth's magnetic field, which turns into an *induced athwartship magnetization* (IAM) when the vessel steams west. The ILM and IAM flux patterns and triaxial signature contour plots are shown in Figs. 2.3 and 2.4, respectively. It should be noted that the athwartship signature component directly under the keel of a ship with a uniform ILM is zero. Similarly, the longitudinal and vertical keel-line signatures are zero for a uniform IAM. The signature shapes can be linked to their respective flux patterns drawn in the upper right of the two figures.

In practice, a naval vessel does not magnetize uniformly. The distribution of steel through out the volume of a ship is irregular and, as a result, so is its magnetization. The irregular geometry of the hull (bow shape, superstructure, submarine sail, etc.), along with its external appendages (sonar dome, rudders, shafts, struts, etc.), focuses and concentrates the earth's field in these areas causing the induced magnetization to deviate from the uniform condition. The clumping of magnetic material (machinery, fuel and water tanks, weapons, stores of ferrous materials, etc.) inside the vessel, and the irregular flux paths created by decks, bulkheads,

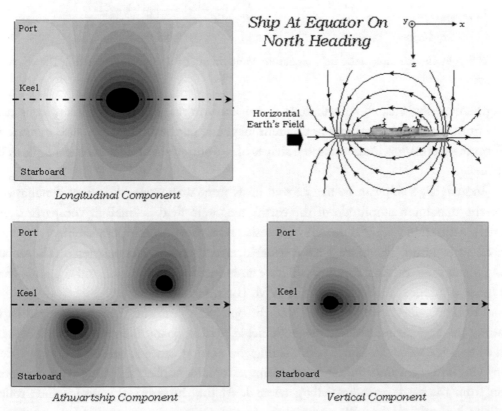

**FIGURE 2.3:** Induced magnetic field signature components of a longitudinally magnetized ship

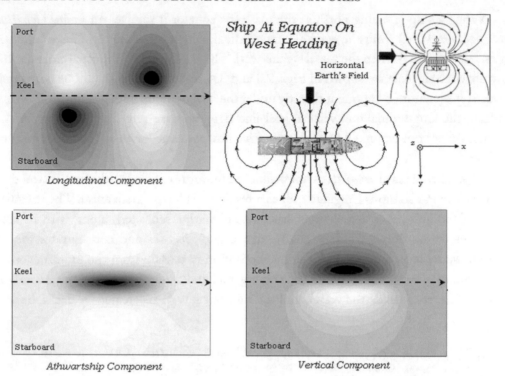

**FIGURE 2.4:** Induced magnetic field signature components of a ship magnetized in the athwartship direction

strengthening ribs, etc., all contribute to a non-uniform magnetization. Finally, constructing a ship from steels with significantly different magnetic permeability would also produce non-uniform magnetization. As a result, the shapes of real ship signatures are not as simple as those shown.

In general, a ship can be magnetized in its three orthogonal directions simultaneously. Since the maximum amplitude of the earth's magnetic field is small in comparison to the magnetic saturation level of the steel's hysteresis curve, the magnetic permeability can generally be taken as constant. Therefore, a naval vessel located at an arbitrary position on the globe and sailing on an inter-cardinal heading will have an induced magnetization and signature that, to first order, is a linear combination of the IVM, ILM, and IAM.

Hysteresis curves for ship steels and the operating point on them are affected by temperature, the application of large magnetic fields, and mechanical stress. As the temperature of steel reduces below its Curie point during the casting process, some of its magnetic domains will be frozen into the direction of any externally applied field that might be present either from the earth or industrially generated. At the ship yard, the steel is then rolled to form the hull, picked up with cranes that use electromagnets; is cut, welded and subjected to stresses in many forms. As a result of the manufacturing and construction processes, ships leave the yard with significant residual or permanent magnetization, which is simply called *perm*.

Like the induced, a ship's permanent magnetization can be broken up into the three orthogonal directions. The three perm components are called the *permanent vertical magnetization* (PVM), *permanent longitudinal magnetization* (PLM), and *permanent athwartship magnetization* (PAM). Each of the three permanent magnetization components generates their own longitudinal, athwartship, and vertical magnetic field signatures. This brings the total number of ferromagnetic signature components up to 18; three each for the ILM, IAM, IVM, PLM, PAM, and PVM.

The triaxial magnetic field signatures of a 13 300 ton commercial steel hull surface ship were measured at a depth of 25 meters below the vessel, and 67 meters horizontal distance from its keel. The longitudinal, athwartship, and vertical signatures were recorded as the vessel sailed by a submerged magnetic field sensor, and are plotted in Fig. 2.5 as a function of time. The earth's background field has been removed from the data. By comparing the field patterns in Fig. 2.5 to those in Figs. 2.2 through 2.4, it is clear that the vessel is magnetized primarily in its longitudinal and vertical directions. The strengths of these signatures are very large, and exhibit a signal-to-noise ratio greater than 40 db.

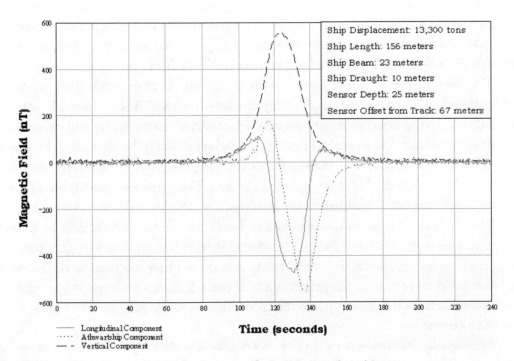

FIGURE 2.5: Triaxial magnetic field signatures of a steel hull surface ship

## 2.3   OTHER IMPORTANT SOURCES OF MAGNETIC FIELD

The second most important contributor to a ship's magnetic field signature is eddy currents. Eddy currents are generated in any electrically conducting material found onboard a ship as it rotates within the earth's magnetic field. This process is similar to what occurs inside an electric generator as its windings cut through the magnetic flux lines established inside it. Eddy currents produce their own magnetic fields that can modify and add to a ship's ferromagnetic signatures.

A conductor does not have to be ferromagnetic to support eddy currents. Ships made from aluminum, stainless steel, or titanium will all have eddy currents induced in them as they rotate. Although in principle eddy currents are produced when a vessel pitches or changes heading, these components are small in comparison to roll induced currents. The reason is that the amplitude of the induced voltage and current is proportional to the angular rotation rate, which is largest in the roll direction. The roll period of some smaller surface ships and metallic boats can be as short as 3 seconds in duration.

Under certain sailing conditions, eddy current generated field signatures can be on the same order as the ferromagnetic components. In addition, the frequency content of eddy currents (ship roll period) falls in the same ULF band used by the threat systems to detect a ship's ferromagnetic fields. Therefore, from a ship protection viewpoint eddy current signatures are very important. However, in calm waters and at moderate speeds no earth induced eddy currents are generated in a vessel. So from a mine or surveillance system designer's viewpoint, the generation of eddy current signatures by a target ship will aid in its detection, but are too unreliable to be deliberately exploited as a primary influence field.

The third largest, and least known of the major sources of magnetic field, is corrosion currents that flow in and around a surface ship or submarine's hull. When a vessel's steel hull is electrically connected to its bronze propeller and immersed in seawater, a battery is formed. The primary path for corrosion currents is from the ship's hull through the seawater to its propeller (some ships have more than one), then up the shaft through the bearings and drive mechanism, and eventually back to the hull to complete the circuit. The corrosion currents are a source of both static and alternating magnetic field signatures.

Electric current flows in seawater by a different mechanism than it does in a metallic conductor. Pure water is a good dielectric insulator. Water's hydrogen and oxygen atoms form strong covalent bonds, and under normal conditions, do not have electrons in the conduction band that could carry electric charge through it. If enough energy (voltage) is applied to pure water, some electrons will move up into the conduction band allowing a current to flow in the form of an electric arc.

Mixing salt into pure water will produce a conducting electrolyte. Hydrogen atoms bond asymmetrically to oxygen producing a polarized water molecule that is strong enough to pull salt atoms apart. Ionized sodium chloride constitutes 88% of the salt in seawater. The positive sodium ion, called a *cation*, is attached to the negative side of the water molecule, while the negative chlorine ion, called the *anion*, has the positive hydrogen side fixed to it. (It should be noted that the water molecules do not form a chemical bond with the ions.) When a voltage is applied to electrodes placed in seawater, the positively charged cation flows toward the negative electrode called the *cathode*, while the negative anion flows to the positive electrode called the *anode*. Unlike a metal conductor, electric current in seawater is comprised of both positive and negative charges that flow past each other toward anodes and cathodes of opposite polarity.

The primary driving voltage that produces corrosion currents in a ship is the electrochemical potential difference between its steel hull and nickel-aluminum-bronze (NAB) propeller. Typically, electrochemical potentials of various materials are reported referenced to a silver-silver chloride electrode, which for steel is around −650 mV and for NAB approximately −230 mV [10]. Although the 420 mV potential difference between steel and NAB seems small, the large surface area of the hull and its high conductivity result in sizable corrosion currents and magnetic fields.

Cathodic protection systems are used to prevent a ship's metallic hull from corroding. The principle of their operation is to turn anodic materials into cathodes. There are two types of cathodic protection systems in use today on naval vessels. The first type, called a passive cathodic protection system, is comprised of a large number of zinc bars that are welded to the hull. Since the electrochemical potential of zinc is near −1000 mV, it becomes an anode when attached to the hull, which is turned into a cathode and protected from rusting. Naturally, the zinc bars themselves will corrode and must be replaced periodically. For this reason, the passive method of cathodic protection is sometimes called a *sacrificial anode* system.

A second type of corrosion protection called impressed current cathodic protection (ICCP) is used on large ships. Instead of zinc bars, ICCP system anodes are made of platinum coated wires or rods that are mounted on the hull inside an insulated housing. The anodes are wired to power supplies located internally in the hull that actively pump current into the seawater, turning the hull cathodic once again. The voltage at the ICCP anodes must be constantly regulated to ensure that sufficient current is flowing to protect the ship from corroding, while preventing excessive current from entering the hull that might cause hydrogen embrittlement thereby weakening it. Silver-silver chloride electrodes called *reference cells* are mounted at several positions on the hull to monitor the effects of the anode current and to regulate it accordingly. A ship's ICCP system automatically adjusts its anode currents until the reference cells measure a specified potential relative to the hull, called the *set potential*. Generally, the set potentials for naval ICCP systems range from about −800 to −850 mV with respect to the hull.

Cathodic protection systems, and especially ICCP types, can drive large amounts of current into the sea that then flow through the hull, primarily along its longitudinal axis. (The

ICCP power supplies on board aircraft carriers have a combined capacity in excess of 1800 amperes.) The hull and propeller shaft current are the main sources of offboard corrosion related magnetic (CRM) signatures. The ship can be represented as a DC and AC longitudinal electric dipole with its CRM fields circulating around it according to the right-hand rule.

The current flowing in the seawater does produce a small magnetic field owing to the sea surface and sea bottom conductivity discontinuities. Seawater conductivities range from near zero in fresh water estuaries to almost 6 S/m in hot areas of the ocean. The effective sea bottom conductivities near the interface are generally 1% to 10% than that of water, while the air is of course non-conducting. The jumps in corrosion current density across the two interfaces generate a small field that subtracts from the ship's athwartship CRM signature, simultaneously generating a longitudinal magnetic field component that would not otherwise exist. The vertical CRM signature is not affected at all by the sea floor or sea surface at DC and quasi-static frequencies.

The static CRM signatures fall off with distance at a slower rate than the ferromagnetic or eddy current fields. The reason is that the source of the CRM field is an electric dipole, where the latter two components are produced by magnetic dipoles. This is an important distinction between the sources. At some distance from the vessel, the CRM field may be the only signature component that can be detected. The range at which the CRM field begins to dominate over the other two depends on their relative source strengths. If the magnetic dipole sources are large, then the CRM field is only important at long ranges, assuming that the signal-to-noise ratio of the sensor system is sufficient to detect it. On the other hand, if the ferromagnetic and eddy current sources are significantly reduced or compensated, then the CRM field may dominate the signature even at very short distances important to mine warfare.

Presently, there are two important shipboard sources of corrosion-related alternating magnetic fields. As the shaft rotates, the variable contact resistance between it and the bearings modulates the corrosion currents. The shaft modulated currents excite alternating CRM fields that occur at the fundamental shaft rotation frequency, plus harmonics. In addition, insufficient filtering of the ICCP system's output can impress AC ripple currents into the hull and surrounding seawater. The alternating CRM signatures produced by ICCP ripple current will have components at the ship's electric power frequency or supply-switching frequency depending on the system's design.

Alternating magnetic fields are easier to detect because of better sensor sensitivities at the higher frequencies, along with lower background noise levels in these bands. However, the ICCP ripple can be eliminated with proper filter design, while the shaft related alternating CRM signatures can be significantly reduced with an *active shaft grounding* system [11]. Therefore, exploitation of the alternating magnetic field signature falls into the same category as eddy currents; do not rely on them.

The last of the major shipboard sources of magnetic field are called *stray fields*. Stray field signatures can be produced by any current carrying electric circuit found onboard a ship. The larger stray fields are produced by the electro-mechanical machinery and power distribution system of the vessel. High power electric generators, motors, switchgears, breakers, and the distribution cables that interconnect them can emit both DC and AC fields. Ironically, a ferromagnetic steel hull shields the internal stray field sources to some degree, attenuating their signatures, especially at higher frequencies. Therefore, at present, stray field magnetic signatures are the smallest of the four major sources.

The magnitude and importance of magnetic stray field signatures will increase in the near future. Trends toward constructing naval vessels from non-magnetic metals, such as aluminum or stainless steel, could significantly reduce the DC shielding effectiveness of the hull. In fact, manufacturing ships from non-conducting composites would completely eliminate all the stray field attenuation provided presently by a ferrous hull. Also, the U.S. navy has committed to developing an "all electric" ship that will use large electric motors for propulsion. Since the power supplied to electric propulsion motors could exceed 30 MW, very high voltages, and more importantly, very large currents would be flowing inside the ship's power system. The problem would be exacerbated if the motors are mounted exterior to the ferrous hull where no shielding at all would be present. Both the DC and AC stray field signature components must be combined with the other three sources in assessing a vessel's true susceptibility to magnetic field detection.

There are many techniques to reduce or compensate the four major shipboard sources of the magnetic field. The first rule in any signature reduction design is to eliminate as many sources as are technically feasible and financially affordable, before attempting to actively cancel the remaining fields. For example, constructing naval vessels from non-magnetic and non-conducting materials could yield a 40 db reduction in its total magnetic field signature due to ferromagnetic, eddy current, and CRM sources. (Some ferrous steel that is part of machinery items, weapons, and other ship systems may have to be retained for them to operate properly.) In addition, care in the up-front design of high power electric propulsion motors, generators, and their interconnecting distribution system could trim down a large portion of the stray magnetic fields cheaply and with little impact on the ship. Any magnetic signatures that remain after the source elimination process could be reduced by 20 to 40 db through active field cancellation techniques. A detailed description of the methods to reduce the magnetic field signatures of naval vessels is outside the scope of this discussion.

## REFERENCES

[1]  L. Metcalfe, Encyclopedia: La Gloire. NationMaster.com. Rapid Intelligence Pty Ltd. Sydney, Australia Jan. 2005, [Online]. Available: http://www.nationmaster.com/encyclopedia/La-Gloire.

[2]    ——Encyclopedia: HMS Warrior (1860). NationMaster.com. Rapid Intelligence Pty Ltd. Sydney, Australia July 2005, [Online]. Available: http://www.nationmaster.com/encyclopedia/HMS-Warrior-%281860%29.

[3]    ——Encyclopedia: CSS Virginia. NationMaster.com. Rapid Intelligence Pty Ltd. Sydney, Australia July 2005, [Online]. Available: http://www.nationmaster.com/encyclopedia/CSS-Virginia.

[4]    ——Encyclopedia: USS Monitor. NationMaster.com. Rapid Intelligence Pty Ltd. Sydney, Australia July 2005, [Online]. Available: http://www.nationmaster.com/encyclopedia/USS-Monitor.

[5]    P. A. Davidson, *An Introduction to Magnetohydrodynamics*, Cambridge, United Kingdom: Cambridge University Press, 2001, pp. 166–203.

[6]    S. Mclean, "The world magnetic model." National Geophysical Data Center. Bolder, CO. Dec. 2004, [Online]. Available: http://www.ngdc.noaa.gov/seg/WMM/DoDWMM.shtml.

[7]    R. P. Feynman, R. B. Leighton, and S. Matthew, *Lectures on Physics*, 1st ed., vol. II, Reading, MA: Addison-Wesley Company, 1977, pp. 13–6 to 13–12.

[8]    R. P. Feynman, R. B. Leighton, and S. Matthew, *Lectures on Physics*, 1st ed., Vol. II, Reading, MA: Addison-Wesley Company, 1977, pp. 14–7 to 14–8.

[9]    R. M. Bozorth, *Ferromagnetism*, New York: IEEE Press, 1993.

[10]   H. P. Hack, "Atlas of polarization diagrams for naval materials in seawater", Naval Surface Warfare Center, West Bethesda, MD. Tech. Rpt. TR-61-94/44, Apr. 1995.

[11]   W. R. Davis, "Active shaft grounding." W. R. Davis Engineering Ltd. Ottawa, Canada. 2004 [Online]. Available: http://www.davis-eng.on.ca/asg.htm.

CHAPTER 3

# Exploitation of Magnetic Signatures by Naval Mines

## 3.1 EVOLUTION OF THE MAGNETIC MINE

The first naval mine was invented by David Bushnell, and was deployed against the British fleet by the colonialist during the American Revolutionary War. A tar covered beer keg, with additional wood floatation on its ends, served as the mine's case. These mines were filled with gunpowder and floated in the river towards anchored warships. A flintlock internal to the mine ignited the gunpowder when subjected to a light shock, hopefully produced by contacting a ship's hull. Mines of this type are called *contact mines* since they detonate on contact with the target. Bushnell mines were unreliable due to their touchy firing mechanism, wet gunpowder, and getting stuck upstream from their target [1], and early herald of the cardinal importance of reliability in naval mining.

The development of contact mines proceeded over the years and were deployed during the American Civil War, the Russian-Japanese War, and used extensively through out both World War I and II. The firing mechanism also evolved during this time, resulting in the very reliable chemical horn and electrode firing device. Several of these horns were mounted around a spherical mine case, that was moored by an anchor a few feet below the water surface. The horns were made of soft lead that covered a glass vile filled with an electrolyte [2]. When a ship bumped against one of the horns, the glass inside would break allowing the electrolyte to flow between the two contacts, closing the firing circuit and detonating the mine. During World War I, thousands of chemical horn and electrode mines were moored throughout the waters of Europe and the North Sea. The electrode mines were especially effective as a barrier against submarines [3]. Moored contact mines are still used today. In 1991, during Operation Desert Storm, the *USS TRIPOLI* struck an Iraqi contact mine and suffered heavy damage [4].

Cable-cutting countermeasures to moored contact mines were so effective during World War I that in 1920 Germany started development of mines that actuated on a ship's influence fields. Taking note of the British efforts to develop a magnetic mine, Germany perfected a *bottom influence mine* that sat on the sea floor immune to cable cutters and exploded when it detected a ship's magnetic field. By 1925 German magnetic influence mines were operational, which they used extensively during World War II along the coastline and estuaries of Britain, and later to defend against the allied invasion of France.

The type of field sensor used in the German magnetic mines is sometimes referred to as a *dip needle*. A dip needle operates fundamentally like a pocket compass turned on its side. When a magnet, or in this case a magnetized ship, moves above the magnetized needle, it will rotate or "dip" up or down depending on its relative polarities. When a dip needle is used as a sensor to fire a mine, it must be leveled at startup. This was accomplished in the German magnetic mines with a complicated but masterful set of mechanical gimbals and springs. Tension in the springs was adjusted so that the needle lay horizontally in the local earth's field when deployed.

During World War II the British recovered and examined several German magnetic bottom mines that were laid by aircraft in water that was too shallow. At low tide the mines were exposed, then disarmed and pulled to shore by British ordnance disposal teams. A photograph of a German magnetic bottom mine recovered in this way is shown in Fig. 3.1, along with a simplified schematic of its firing circuit [5]. When the mine enters the water, a pressure safety

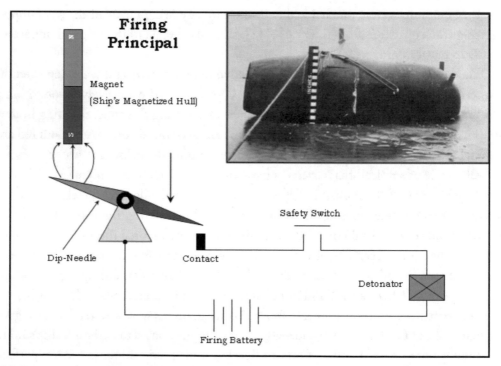

**FIGURE 3.1:** German World War II magnetic bottom mine and schematic of its dip needle firing circuit

switch is closed arming the weapon. As a magnetized ship sails over the mine, the dip needle will rotate until it closes a contact that electrically detonates the mine.

The dip needle magnetic mine could be reconfigured to produce different firing characteristics. The contact could be positioned so that the mine fires on the target ship's north magnetic pole, or its south. With two contact switches, a polarity reversal could be required before detonation, a positive trigger followed by a negative or visa versa. These variations in the mine's firing logic were necessary as the British learned how the mines operated and developed countermeasures to them.

German World War II mines actuated only on a ship's vertical magnetic field signature component owing to the dip needle's orientation. The minimum magnetic field strength needed to fire the mine could be preset by adjusting the mechanics of the firing device. The selection of actuation thresholds available in the German MDA, JDA, and M1 through M5 magnetic mines were 3000 nT, 2000 nT, 1000 nT, 500 nT, and 250 nT. Many factors had to be considered while choosing an actuation level including the objective of the minefield, the size of the target and its expected signature level, the size of the explosive charge and the target's vulnerability to shock damage, and the desired resistance to countermeasures.

The U.S. took a different approach to magnetic mine sensor design. Instead of using a dip needle, an inductive loop served as the magnetic sensor of the mine's *target detection device* (TDD). The voltage at the terminals of an induction sensor, called a *search coil*, is proportional to the rate-of-change in magnetic flux that passes through it. The search coil TDD was made by winding several hundred turns of copper wire over a permalloy rod to increase its sensitivity. An example of a magnetic induction mine is the U.S. Mk 52 [6] shown in Fig. 3.2, along with a simplified schematic of an induction mine's firing circuit [7]. During the passage of a ship enough voltage is generated at the induction sensor's output to drive a sensitive secondary relay, which closed and fired the explosive. Germany's strategic shortage of copper and nickel prevented their production of search coil induction mines, forcing them instead to use the mechanically more demanding dip needle sensor.

## 3.2    MODERN MAGNETIC MINE TECHNOLOGIES

Since World War II, naval mines have improved significantly. (A historical record of the U.S. naval mine development has been documented in *The Legacy of the White Oak Laboratory* [8].) Some of the augmented capabilities of modern mines include:

1. increased lethality and damage radius provided by more powerful explosives;

2. self propelled versions with larger attack ranges that increase the threat level of a minefield while requiring fewer mines to be deployed;

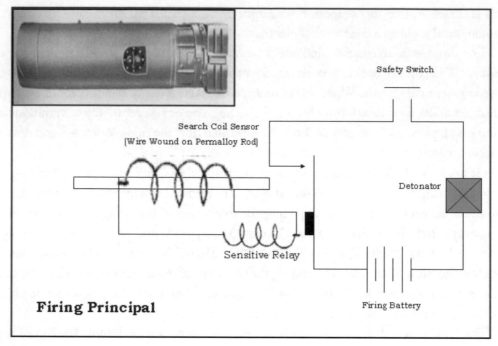

**FIGURE 3.2:** United States Mk 52 magnetic bottom mine and schematic of its induction firing circuit

3. better classification of target types and ship classes; and

4. increased resistance to countermeasures such as mine hunting, mine sweeping, and target ship signature reduction.

Larger attack and damage radii against targets with reduced signatures required a sensitive, low power, inexpensive and rugged magnetic field sensor. Also, greater target detection ranges force the use of a stable low frequency sensor, while target classification and sweep rejection could be enhanced with a small three-axis (triaxial) version. To satisfy these and other unique operational requirements, mine designers turned to fluxgate magnetometers.

Invented in 1931 by H.P. Thomas, a fluxgate magnetometer is a vector magnetic field sensor that operates on a principle similar to a magnetic modulator and a magnetic amplifier [9]. Although there are many fluxgate designs with different performance characteristics, all sensors of this type work on the principle of modulating the quasi-DC signal field inside a cyclically saturated ferromagnetic core. In its simplest form, the fluxgate core is a single small ferrite rod wrapped with a solenoid, called the *drive winding*, and a second smaller pick-up coil (*sense winding*) also wrapped on the core. If the drive winding is supplied with sufficient AC current, the core will saturate alternately in its positive and negative directions, following the ferrite's hysteresis curve. If no external field is present in the core, the sense winding has an output voltage that contains only odd harmonics of the drive frequency. When a signal field, like that from a ship, is added to the core, the hysteresis curve is offset slightly along its abscissa, generating even harmonics with amplitudes proportional to that of the signal. By passing the second harmonic output of the sense winding through a synchronous electronic detector and calibrating its output, both the amplitude and polarity of the signal field can be accurately measured [10].

The push to detect a vessel's magnetic signature at increasingly larger distances has resulted in the development of stable low-noise fluxgates with high sensitivity from DC to a few kilohertz. Sensor components important to achieve these operating requirements include [11]:

1. core material with high permeability, small area inside its hysteresis curve, low magnetostriction, low Barkausen noise, uniform cross section with homogeneous magnetic and mechanical properties, and high electrical resistivity;

2. core geometries and winding configurations such as single-rod, double-rod, ring-core, and race-track with good thermal stability;

3. internal and external field compensation techniques to maintain high sensor linearity; and

4. drive and sensing electronics that are small, low power, low noise, stable, linear over a wideband, and inexpensive.

Additional issues in multi-axis applications are crosstalk, mechanical alignment accuracy, and stability. The numbers of parameters that must be taken into account in the design of a fluxgate magnetometer make it a true art form.

Fluxgate magnetometers that meet the TDD requirements of a mine have existed for many years. Inexpensive compact sensors are available with peak-to-peak noise levels below 0.25 nT in the band from 0.05 to 1 Hz, over the temperature range from $-40^0$ to $+50^0$ C, and consumes only 5 mW of power. These sensors are more than sufficient to detect a steel hull target with a high signal-to-noise ratio out to the maximum damage ranges of modern explode-in-place mines. However underwater surveillance systems, and mines that are incorporated into them, require magnetometers that are much lower in noise over a much broader bandwidth. These sensors will be discussed in the next chapter.

Naval mines use magnetic field detection in combination with other types of influence sensors (acoustic, pressure, seismic, etc.) to fulfill four main objectives. The influence sea mine must:

1. eliminate or significantly reduce ambient natural or manmade background noise;

2. detect the presence of its primary target, such as a class of surface ship or submarine sailing within its attack range, and measure its signatures;

3.   recognize mine countermeasure signals as false targets and inhibit from firing; and

4.   identify those signatures coming from valid targets and to detonate its explosive warhead at a time that meets its lethality requirement.

These objectives are to be satisfied in a statistical sense when the mine is planted in a field (minefield) with other similar weapons. To accomplish these objectives, naval mines combine the outputs from one or more sensors that detect variations in hydrodynamic pressure, wideband acoustic/seismic signals, magnetic fields and their gradients, and/or other influences. Signal processing of signature characteristics from one or more sensors can improve the mine's performance in meeting its four objectives.

When two magnetic sensors are mounted on a rigid base, with their sensing axis aligned and their output signal subtracted, a magnetic *gradiometer* is formed Gradiometers detect near-by sources of field that typically have high gradients, but are less sensitive to distant sources that produce uniform fields over the length of the instrument, which are then subtracted out. As a result, mines that employ magnetic gradiometers are less susceptible to distant noise sources, false targets, and mine motion and can be more selective in detonating only when a vessel is within its lethal range.

The use of a pressure sensor in a mine is a two-edged sword. The pressure signature of a ship is produced by the Bernoulli Effect of water flowing from its bow to the stern [12]. Although it is difficult to artificially mimic this influence signature with mine sweeping systems, it has drawbacks when used. The low frequency response of a pressure sensor is limited by the much larger hydrostatic pressure at depth due to dynamic range issues. This makes the mine susceptible to missing slow moving vessels within its lethal range. Also, surface waves and swells can produce pressure variations (background noise) in the sensor's passband, masking the target's signature or causing the pressure channel to be constantly activated. Incorrect settings for pressure mines run the risk of a catastrophic failure of the minefield, attacking no targets within their damage range.

When used in combination with magnetic sensors, acoustic fields confirm the presence of a target. Acoustic fields propagate very efficiently in the ocean, and can be detected at large distances from their source, a significant disadvantage for mines with a limited attack range. A ship's acoustic signature is generated by machinery, vibrating plates and ship structure, propellers, and turbulent flow of water around the hull. Combination magnetic-acoustic or magnetic-seismic mines are less susceptible to premature detonations by background noise and distant targets, and also complicate mine sweeping operations.

Although older mines use analog circuitry in their TDD to decide whether or not to fire, modern influence mines are controlled by microprocessors, allowing their firing hardware to be programmed to meet many more scenarios with quick and easy changes to the installed software.

These low power controllers monitor the mine's arming and influence sensors to decide if all requirements have been met to activate, and then follow preprogrammed logic to determine if and when it should fire. For example, when an acoustic-magnetic mine's hydrostatic switch has been activated by its water deployment, the mine begins to monitor the background acoustic field. When the acoustic energy in its passband continuously exceeds a preset threshold for a sufficient time, the controller will turn on the magnetic sensor and begin monitoring the output. If the magnetic field threshold is not exceeded within a specified time, the mine will turn off the magnetometer and switch back to monitoring the acoustic channel. If instead the magnetic threshold is exceeded for the required time, the mine will actuate and either advances its ship-counter by one or fire if its ship-counter is at zero. (A ship counter is used to confound mine sweeping operations or to produce a delayed threat.) The roles of the magnetic and acoustic channels can be reversed by first monitoring with the magnetic sensor and actuating on the acoustic.

The firing logic used in the operational naval mines is usually much more complicated than the example. Multiple thresholds, signature polarities, and rate-of-change requirements can be incorporated into the firing logic. Timing and delay functions are extremely important in the programming of a TDD, both within decision modules and in between processes. A microprocessor controlled TDD can even perform complicated real-time correlations between multiple sensor channels to enhance its capabilities in meeting the mine's four main objectives. However, influence mines must continue to operate for up to a year on a single small battery and can not power a high performance processor for very long.

Knowledge of the sensitivity, type, and frequency response of sensors, and the firing logic used in the actual mines have obvious important military consequences. When this information is known, mine countermeasure systems and signature reduction technologies might be tailored to defeat the weapon and eliminate it as a threat. For this reason, specifics of operational mine systems or those in development are kept secret, closely guarded by all nations.

Since an influence mine deployed on the sea floor does not make contact with the target ship when it explodes, the mechanism for imparting damage to its hull may not be obvious. When a mine explodes, a high-pressure gas bubble in excess of 100 000 atmospheres expands against the water, generating a shock wave that travels initially at supersonic speeds [13]. When this pressure pulse impinges on a ship, it will transfer its kinetic energy to the hull that will deform in the process. If the ship's hull can not completely absorb the shock energy through plastic deformation, it will rupture.

The dramatic damage to a ship caused by a non-contact underwater explosion can be observed during full-scale sinking exercises. The Australian destroyer *HMAS TORREN* was sunk with a 650 lb (300 kg) non-contact explosion detonated several feet below its keel. Bending of *TORREN*'s hull by the impinging shock wave is clearly evident in the photograph shown in

**FIGURE 3.3:** Sinking of the Australian destroyer *HMAS TORREN* by a 650 lb. non-contact explosion detonated several feet below its keel

Fig. 3.3(a). As the gas bubble produced by the explosion collapses, its bottom moves up faster than its top goes down. The momentum of the water causes a new secondary bubble to form closer to the hull, imparting a second shock to the hull along with a jet of water (Fig. 3.3(b)). The combination of shocks partially lifts the ship out of the water causing it to flex violently, whipping it in half (Fig. 3.3(c)). The separated aft portion of the destroyer sank quickly leaving just the forward half afloat (Fig. 3.3(d)) [14].

Even if a vessel's hull does not rupture, energy from the pressure pulse will be transmitted throughout its structure as shock and vibration. The rapid acceleration of the ship's structure from a shock wave can injure the crew, severely damage machinery and propulsion systems, weapons, and electronic devices. Therefore, a mine does not have to sink a ship to be effective; disabling it sufficiently to abort its mission may have the same immediate effect on the battle. Depending on the scenario and objective of the minefield, causing a ship to suffer moderate damage may delay it adequately to affect the outcome of the engagement. Even a slight mine-damage to a ship may adversely impact battle plans, or the psychology and political aspects of the conflict.

As discussed in Chapter 2, the magnetic field signatures of naval vessels are highly variable in amplitude and shape making it difficult to program a mine's firing logic. This is also true of other influence fields. As a result, setting a mine's actuation thresholds must be handled on a statistical basis.

Full-scale measurements or mathematical models that capture the variation in the magnetic field signatures of naval vessels can be used to determine the probability of a ship actuating a mine as a function of its relative location. Passing large numbers of signatures through hardware or software simulations of an influence mine will produce a *probability-of-actuation* curve similar to the idealized example shown in Fig. 3.4. The area under the probability-of-actuation curve is called the *average firing width*. The average firing width is sometimes used in minefield analysis as the actuation curve for the mine, and has a probability of 1 inside this width and zero outside. For statistical simulations of random passages of target ships, using the average firing width will usually give essentially the same number of actuations as the probability curve for a specific mine setting [15].

Similar to actuation, a *probability-of-damage* curve and an *average damage width* can be generated for the mine's explosive charge and the statistical variations in the target ship's shock damage. Combining the probability-of-actuation and probability-of-damage curves statistically will give the true vulnerability of target ships to a mine, and can be used to set its optimum actuation threshold (*sensitivity setting*) to achieve the objective of the minefield planner. However, in this discussion, well defined actuation and damage contours will be used in order to more clearly explain the interaction of mine sensitivity settings, mine countermeasures, and magnetic signature reduction.

A mine's sensitivity setting must be selected by the minefield planner to ensure that it actuates on a passing ship so as to inflict the desired degree of damage, simultaneously being

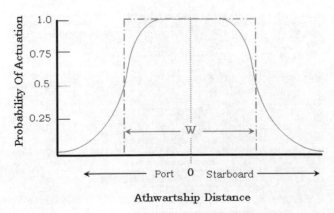

FIGURE 3.4: Example probability-of-actuation curve and its simplified representation as an average firing width

resistant to environmental noise and mine countermeasures. As an example, if the purpose of the minefield is to inflict enough damage on transiting naval vessels so that they can no longer complete their mission (*mission abort damage*), then the optimum setting should cause mine actuation at a range equal to the mission abort damage distance of its charge. Well-defined simplified actuation and damage contours (analogous to average actuation and damage widths) are used in this explanation to remove the statistical vagueness of their boundaries. The actuation contour for an ideal optimum mine setting, along with those that are considered to be suboptimum in this example, are shown in Fig. 3.5. As drawn in the figure, mines that are set too sensitive may detonate on targets that are too far away to inflict on them the desired level of damage called for in the minefield plan. These mines might be considered wasted and will leave a gap in the minefield that may be important to the outcome of the battle. In addition, mines that are too sensitive are easier to sweep, which will be discussed later. Conversely, if the mine is set too insensitive it will miss targets that it could have heavily damaged or even sunk. Setting mines too insensitive could result in a catastrophic failure of the minefield presenting no threat to the transiting vessels. For these reasons, a naval vessel's signature characteristics (amplitude and shape) are also closely guarded.

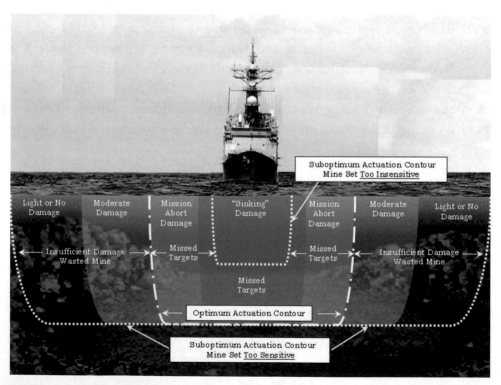

**FIGURE 3.5:** Idealized example of a magnetic mine actuation contour matched to its mission-abort damage contour, along with two settings that are sub-optimum

## 3.3    MAGNETIC MINE COUNTERMEASURES

Mine warfare, and in particular MCM strategies, are complicated. The risk in losing a ship or submarine to a mine is very scenario dependent, and is sensitive to many parameters including:

1. density of mines in the field (number per square kilometer),
2. availability of mine hunting and sweeping platforms in theater and their effectiveness in the specific ocean environment,
3. mission plans and their time constraints,
4. required length and width of the "Q" route (transit lane) and area needed to conduct operations,
5. susceptibility of combatants to actuating a mine during their transit through the field,
6. vulnerability of the vessel to damage from the mine's explosive charge if it donates.

Although the absolute effectiveness of mine-clearing operations and its impact on the overall mission depends highly on the parameters listed, the functional relationship of combatant losses to MCM tactics and technologies has a well-defined trend irrespective of the scenario details.

Historically, active MCM originated to defeat the moored mine threat. During World War I and in the early days of World War II, steel serrated cutting chains draped between two shallow-draft vessels were used to snag and saw through a mine's mooring chain. Later, an explosive cutter was developed that allowed a single ship to grab and separate the mine from its mooring. After the mooring chain is cut, the mine floats to the surface where it can be detonated or incapacitated by gunfire. This technique is called *mechanical mine sweeping*.

Another countermeasure technique employed extensively today is mine hunting. A sonar deployed from a ship, helicopter, or installed on an unmanned underwater vehicle (UUV) can easily detect mines floating in the water column. In addition, recently developed airborne blue-green laser technology can detect weapons moored close to the surface. After a hunting system locates a mine, it can be destroyed with a small explosive, or if time is critical, its location can be marked and simply avoided by ships if the scenario allows. However, in comparison to moored mines, hunting bottom magnetic mines is significantly more difficult.

All good defenses are layered, as is the case in mine warfare. The first and best defense against mines is to prevent their manufacture, transport, and deployment. Owing to the tactical or political constraints, many of them may slip through and be deployed. The second defensive layer, *active mine countermeasures*, involves mine detection through hunting, destruction with explosive charges, and decoying with *influence mine sweeping*. Influence sweeping systems produce signatures that are designed to actuate specific mine systems or that mimic those of combatants. Sweepers attempt to trick the weapon into detonating at safe remote distances. However, one or more live mines may be missed during MCM operations due to mission time constraints, unfavorable environmental conditions, excessive bottom clutter with many mine-like sonar contacts, equipment malfunctions, operator error, or poor planning, etc. The burden then falls on the last layer of ship protection, the underwater signature reduction, to hide a vessel from attack by a mine or to blind it with a jamming signal.

Naval platform susceptibility to bottom influence mines has a parabolic dependence on the amount of MCM effort used before the combatant transits the field. Figure 3.6 shows hypothetical examples of this parabolic relationship for dense, medium-dense, and sparse minefields. The units for MCM effort, platform-days, are equal to the sum of the number of days each MCM platform (ship, helicopter, unmanned underwater and surface vehicle, etc.) is used to hunt, sweep, or otherwise dispose of threat mines. It should be noted that after only a few platform-days of MCM effort, the risk of transiting ships in a dense minefield could be virtually unchanged. Although the absolute scales on the axis of the graph and the relative separation between the three curves will depend on the specific scenario, the trends shown in the figure apply to any minefield.

The MCM effectiveness curves show several important characteristics of mine-clearing operations that can be used to plan a strong defensive strategy. First, the time constraints of a conflict, combined with the availability of MCM resources, will limit the best possible clearing effort to some fixed value. As demonstrated by the vertical line in Fig. 3.6, the risk to combatants

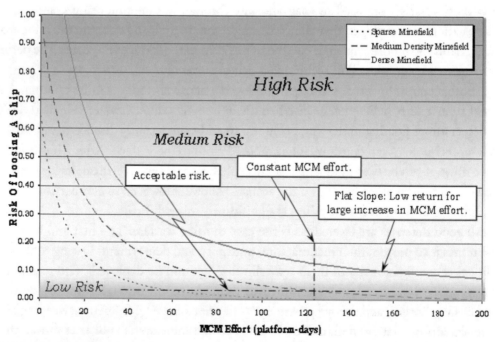

**FIGURE 3.6:** Hypothetical examples of the relationship between mine countermeasure effort and the risk of losing a following ship

would then vary depending on the density of the mines encountered. The first and best MCM strategy is to prevent mines from being laid, or to keep a sparse minefield from becoming dense by denying enemy forces the opportunity to deploy the weapons.

Losing a ship to a minefield or inflicting casualties to its crew is unacceptable. The sinking or occurring mission abort damage by even a single naval platform could lengthen the conflict and also increase the time needed to prepare for the next battle. This is especially true today with the smaller numbers of high-capability ships in modern navies. Therefore, a low to very-low risk level is required for transiting combatants. The intersection of the horizontal line in Fig. 3.6 with the asymptotic portion of the effectiveness curves shows that achieving a low-risk condition could require significant or even unachievable amounts of MCM resources and time, depending on the mine density. Diminishing returns of the MCM effectiveness curves (flattening at higher levels of MCM effort) are caused by the resource-intensive process of removing the last one or two mines from the field; a characteristic of all mine-clearing scenarios. It takes only one missed \$10 000 mine to sink a \$2 000 000 000 ship.

Reducing a ship's underwater magnetic signature can significantly improve the effectiveness of mine hunting and sweeping systems. Underwater ship signature reduction, sometimes called *underwater stealth*, is achieved through elimination of their sources and by active cancellation. Minimizing a naval vessel's underwater signatures makes each of the mine's four objectives (reduce noise, classify the target, reject false targets, and localize the target and detonate) much more difficult, and can actually render it ineffective.

Like stealth aircraft flying against air defense systems, reducing underwater signatures can shorten an influence mine's actuation radius to a point where it is no longer a threat. Therefore, water depths deeper than the influence mine's attack range need not be immediately cleared, with the buffer zones along the edges of transit lanes and maneuvering areas. In addition, vessels employing underwater stealth technologies would have a reduced probability of actuating any residual mines that might have been left in shallower waters after clearing.

Decreasing the attack radii of deployed influence mines is analogous to reducing the effective density of the field. For example, if 100 mines have been deployed in an area but only 25 can detect the transiting targets due to their low signatures, then the effective density of the field has been reduced by 75%. As shown by the parabolic mine-clearing curves in Fig. 3.7, underwater signature mitigation can reduce the effective mine density, improving the efficiency of MCM operations by significantly lowering the time needed to achieve a low risk condition. Eventually, all mines will have to be removed from the field before naval and commercial ships not equipped with stealth technologies can transit the area, but this can be accomplished after the time-constrained forced entry or strike phase of the operation has been completed.

A second way that underwater stealth improves MCM effectiveness is to increase the efficiency of mine sweeping. As discussed previously, the firing thresholds of mines are ideally set so that their detonation results in a high probability of a kill. If ship signatures are reduced and the mine actuation thresholds are not, then their attack range will be smaller, decreasing once again the effective density of the minefield. Depending on the scenario, using too coarse of a sensitivity setting could result in a catastrophic failure of the minefield allowing all ships to pass safely. If, however, the mine actuation thresholds are decreased to reoptimize them with their damage radii, then the more sensitive settings will make them easier to sweep.

There are several types of mine sweeping systems. The "open-tail" or electrode sweep is a single cable pulled behind a surface vessel or a sled towed by a helicopter. The towing vessel or sled is equipped with a high power electric generator that drives hundreds to thousands of amperes through the cable, out one of the electrodes grounded to seawater, and then returns through the second electrode separated by 100 meters or more from the first. For the "closed-loop" sweep, high currents are passed through a large continuous loop of cable that is depressed and widened with underwater Para vanes called "otters". A third magnetic sweep system uses a string of permanent magnets towed through the water behind a shallow draft vessel. Although there are advantages and disadvantages to each of the magnetic sweep systems their purpose is the same, to decoy a mine into actuating on it instead of the following ships.

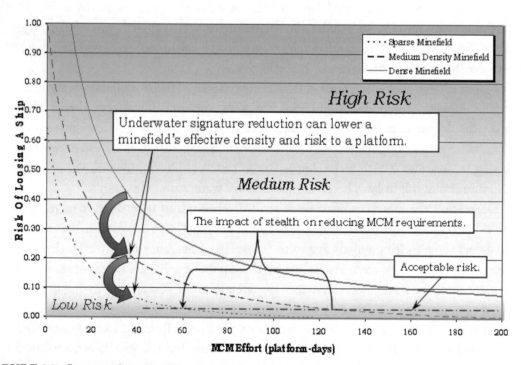

**FIGURE 3.7:** Impact of magnetic signature reduction on the hypothetical example relating mine countermeasure effort and risk to following ships

A direct relationship exists between mine sweeping effectiveness and ship signature reduction. For example, suppose a ship channel is to be seeded with magnetic bottom mines whose actuation sensitivity can be set to one of the 5 previously listed; 3000 nT, 2000 nT, 1000 nT, 500 nT, and 250 nT. If the objective of the mine field is to destroy or disable all commercial container shipping with a vertical signature similar to the one shown in Chapter 2, then the optimum mine setting should be 3000 nT (Setting #1). It is assumed that the damage range of the mine's explosive charge against the container ship is equal to the distance at which the ship's peak vertical field reaches 3000 nT. The mine's actuation setting and mission abort damage range are matched and optimized.

If, in the example, it were known that the ship's signature had been reduced, then one of the more sensitive actuation settings would have to be used instead. Reducing the container ship's vertical signature from 3000 nT down to 2000 nT at the mine's explosive damage radius will force the use of Setting #2 instead of #1. If for whatever reason this change in setting is not made, then the ship's 2000 nT signature will not exceed the mine's 3000 nT actuation level, and the risk of a catastrophic failure of the entire minefield exists. This argument can be extended to reduced ship signature and mine actuation levels of 1000 nT, 500 nT, and 250 nT as charted in Fig. 3.8.

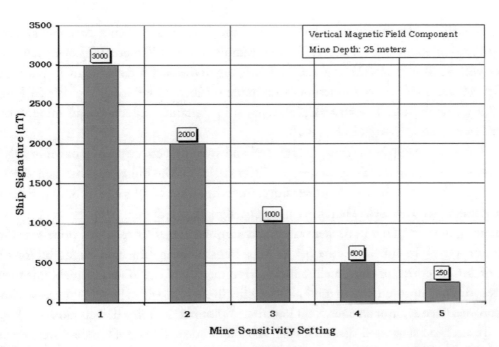

**FIGURE 3.8:** Relationship between ship signature levels and optimum mine sensitivity settings

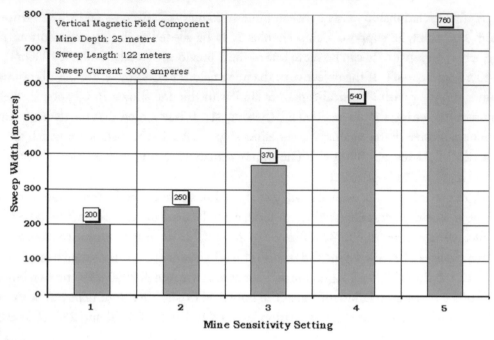

**FIGURE 3.9:** Relationship between swept widths and mine sensitivity settings

As a mine's sensitivity setting is increased the distance at which a sweeping system can decoy it into actuating also increases. If in the example above a 120 meter long open-tail sweep is driven with 3000 amperes, then the *swept widths* (sum of the port and starboard distances swept) of Fig. 3.9 are obtained as a function of the mine settings. Combining the data in Figs. 3.8 and 3.9 yields the direct relationship between ship signature reduction and mine sweeping effectiveness, and is charted in Fig. 3.10.

The relationship between signature level and sweeping effectiveness shown in Fig. 3.10 was computed for the mine's optimum sensitivity setting. If the mine is set less sensitive it will miss targets. If instead it is set more sensitive, then the swept widths in Fig. 3.10 will be larger for a given signature level. Therefore, magnetic signature reduction can reduce the demand for hunting bottom mines in deeper waters and improve the effectiveness of mine sweeping in shallower waters. In either case the risk to ships transiting the minefield is reduced for a fixed level of MCM effort, or the timeline for clearing the minefield to an acceptable risk level is shortened. The improvements in mine clearing effectiveness provided by signature reduction can be expressed once again as an equivalent lowering of the effective mine density curves in Fig. 3.7.

It has been suggested that increasing the amplitudes of a vessel's underwater signatures would reduce its risk to influence mines by detonating them in front of the ship, beyond the weapons' damage range. However, the firing logic found in modern multi-influence weapons

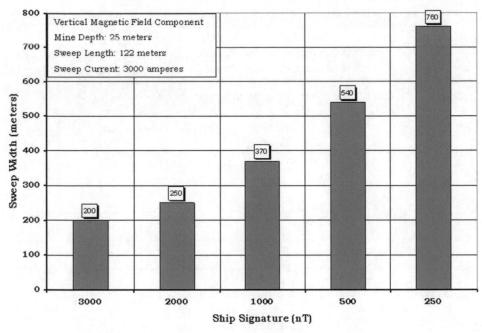

**FIGURE 3.10:**  Relationship between ship signature levels and swept widths

prevents this from occurring. Instead, deliberate signature amplification would raise the effective density of the field by increasing the actuation ranges and threat from those mines that were previously rendered ineffective against stealthy combatants, while providing little protection to follow-on traffic (Fig. 3.11). Risking a $2 000 000 000 manned combatant to sweep a minefield instead of a helicopter sled or unmanned surface vehicle is not a good strategy.

Ironically, a submarine, the quintessential stealthy naval vessel, can not use all the mine-clearing tools available to the surface ships. To remain undetected, precursor sweeping before transiting a minefield is generally not an option for submarines. Even if unmanned underwater sweep systems were available, their use and subsequent detonation of mines would immediately give away a submarine's approximate location, or reveal its intended lane of transit. A submarine must rely solely on hunting mines, avoiding them, and covert (non-detonating) neutralization.

Removing sweeping from the mine-clearing toolbox raises the minefield's effective density. For reasons discussed, all mines may not be detected during hunting operations. In addition, the loss of clearing efficiencies that would have been realized through sweeping increases the overall amount of mine hunting effort (platform-days) necessary to reduce the vessel's risk to an acceptable level. Therefore, more mines will remain in the field (higher effective density) for a mine-hunting-only scenario than an equivalent case that includes sweeping (Fig. 3.12). As a result, a submarine, or for that matter a surface ship requires more underwater stealth and levels of signature reduction if mine sweeping is not used.

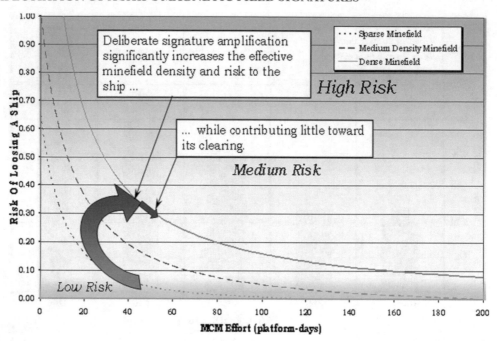

**FIGURE 3.11:** Why deliberate signature amplification is a bad idea for clearing minefields

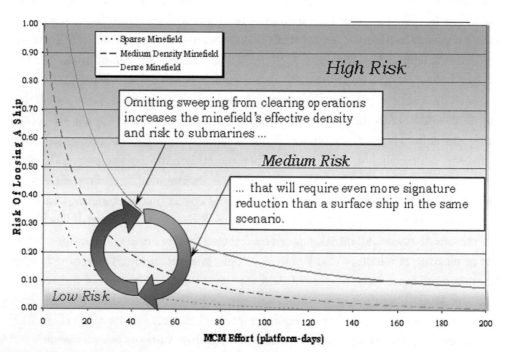

**FIGURE 3.12:** Impact of omitting mine sweeping from clearing operations

As the signal-to-noise ratio of underwater influence fields decreases through the application of stealth technologies, the mine's signal processing and actuation logic must also be changed to prevent degradation of its effectiveness. To this end, modern mine designs have more complex logic and decision trees incorporated into their firing solution. However, each decision point placed into a mine's firing logic is a potential vulnerability to an emerging countermeasure technique called *mine-jamming*.

The purpose of mine-jamming is to fool the mine's logic and prevent it from coming to a correct firing decision while a valid target vessel sails past it. The jamming signal can be generated by either an onboard or offboard field source. The temporal and spatial characteristics of the jamming signal are selected to cause one or more of the mine's decision points to reject the signal as either noise or coming from a decoy or mine sweeper. Therefore, as the underwater signatures of naval vessels are reduced, unsophisticated mines are more easily swept or become ineffective, while modern, sensitive and sweep-resistant mines are easier to jam.

# REFERENCES

[1] S. Underwood, "The first mine: Bushnell's keg." Mobile Mine Assembly Group, Corpus Christi, TX June 2005, [Online]. Available: http://www.cmwc.navy.mil/COMOMAG/Mine%20History/Bushnell%20Keg.aspx.

[2] R. Hoole, "The development of naval mine warfare." The Mine Warfare & Clearance Diving Officers Association, Fareham, United Kingdom 2002. [Online]. Available: http://www.mcdoa.org.uk/MCD_History_Frames.htm.

[3] G. K. Hartmann and S. C. Truver, *Weapons That Wait*, 2nd ed, Annapolis, MD: Naval Institute Press, 1991, pp. 42–55.

[4] P. R. Yarnall, NavSource online: Amphibious photo archive. NavSource Naval History. Baytown, TX Sept. 2005, [Online]. Available: http://www.navsource.org/archives/.

[5] R. Hoole, "HMS VERNON-Before the excavators came." The Mine Warfare & Clearance Diving Officers Association, Fareham, United Kingdom 2002. [Online]. Available: http://www.mcdoa.org.uk/HMS_Vernon_Master_Page_Frames.htm.

[6] T. DiGiulian, "United States of America mines. Welcome to NavWeaps," May 2005, [Online]. Available: http://www.navweaps.com/Weapons/WAMUS_Mines.htm.

[7] G. K. Hartmann and S. C. Truver, *Weapons That Wait*, 2nd ed, Annapolis, MD: Naval Institute Press, 1991, pp. 63–64.

[8] W. B. Anspacher, B. H. Gay, D. E. Marlowe, P. B. Morgan, and S. H. Raff, *The Legacy of the White Oak Laboratory*, Dahlgren, VA: Naval Surface Warfare Center, 2000, pp. 1–23.

[9] P. Ripka, *Magnetic Sensors and Magnetometers*, Boston, MA: Artech House, 2001, p. 75.

[10] D. I. Gordon and R. E. Brown, "Recent advances in fluxgate magnetometry," *IEEE Trans. Mag.*, vol. 8, no. 1, Mar. 1972.

[11]  P. Ripka, *Magnetic Sensors and Magnetometers*, Boston, MA: Artech House, 2001, p. 79–120.

[12]  G. K. Hartmann and S. C. Truver, *Weapons That Wait*, 2nd ed, Annapolis, MD: Naval Institute Press, 1991, pp. 91–93.

[13]  G. K. Hartmann and S. C. Truver, *Weapons That Wait*, 2nd ed, Annapolis, MD: Naval Institute Press, 1991, pp. 98–101.

[14]  S. Connolly and S. Farrow, Mark-48 torpedo war-shot. Naval Sea Systems Command, Washington, DC Sept. 2005, [Online]. Available: http://www.dcfp.navy.mil/mc/presentations/Mark-48.htm.

[15]  G. K. Hartmann and S. C. Truver, *Weapons That Wait*, 2nd ed, Annapolis, MD: Naval Institute Press, 1991, pp. 101–102.

CHAPTER 4

# Exploitation of Magnetic Signatures by Submarine Surveillance Systems

## 4.1    EVOLUTION OF THE SUBMARINE MAGNETIC DETECTION SYSTEM

The development of technologies to detect submarines with active and passive sonar began with World War I, which at the time was named ASDIC after its design organization, the Anti-Submarine Detection Investigation Committee. Sonar technology advanced during World War II, where it was used extensively, and its development accelerated considerably during the ensuing Cold War. The spherical spreading of submarine acoustic fields and their channeling by gradients in sound velocity found in deep waters allowed their characteristic signatures to be detected at distances exceeding hundreds of miles. However, excessive noise and reverberation found in shallow water and harbors limited acoustics to much shorter ranges than in the open ocean.

During World War II, anti-submarine acoustic and magnetic field sensing systems were deployed outside the entrances to harbors, within large bays, and other shallow water areas of military importance. These detection arrays were monitored continuously on shore by operators manually analyzing the strip chart recordings and correlating the contacts on underwater systems to surface radar or visual sightings of surface ships. When an underwater contact did not correlate with the presence of a surface ship, a military vessel was dispatched to prosecute the intruder. In some cases, a line of predeployed mines controlled from the shore by underwater electric cables, were commanded to detonate in the vicinity of the contact.

The World War II underwater magnetic barrier was comprised of one or more large induction loops, sometimes called *indicator loops* or *harbor loops*, deployed horizontally on the sea floor. The loops were several kilometers long and usually less than 0.5 km wide. When a magnetized submarine sailed over the loop, a voltage was induced in it proportional to the rate of change in the magnetic field linking the loop. The signal was cabled to the shore where it was amplified, integrated with a fluxmeter, and recorded on a strip chart. In some cases the signal was played over a loudspeaker.

Over 50 Allied harbors and estuaries around the world were protected during World War II by magnetic induction loops, the most famous being the six loops installed to protect Sydney Harbor, Australia. On the night of May 31, 1942, the Japanese launched a midget submarine attack against the vessels moored inside Sydney Harbor. Although the submarines were detected by the harbor loops and some were destroyed before they could complete their mission, slow response to the contacts and one inoperable loop resulted in the sinking of several Allied ships [1]. A more complete historical description of World War II indicator loops can be found in [2].

Today, magnetic induction loop barriers could find applications in homeland security, force and harbor protection, drug interdiction, and monitoring of coastlines. In addition, arrays of small low-power and inexpensive portable magnetic field sensors could be deployed on the sea floor or in disposable buoys to monitor large portions of the oceans as protection against acoustically quiet hostile submarines. Sensitive magnetic field sensors can be installed on manned and unmanned mobile underwater and surfaced surveillance platforms that can hunt for the slow moving or bottomed diesel boats. Ultimately, magnetic anomaly detection (MAD) of quiet submarines with fast moving airborne platforms may yield the highest search rate.

Magnetic detection of submarines from maritime patrol aircraft equipped with MAD sensors also started during World War II. In June 1942, Project Sail was started under the direction of Naval Ordnance Laboratory to develop and test MAD systems that employed fluxgate magnetometers as airborne sensors. These systems were installed on VP-63 Catalina aircraft. On February 24, 1944, in the Straits of Gibraltar, VP-63 became the first aircraft to magnetically detect a submarine that resulted in the sinking of *U*-761. The MAD aircraft patrols were so successful in stopping submarine traffic through the straits that Sir Andrew Cunningham, Admiral of the Royal Navy, recognized the VP-63 squadron as turning the Mediterranean "into an Allied lake" [3].

During the ensuing Cold War years, advances in passive and active acoustic submarine detection relegated the use of MAD to become only a localization system for releasing weapons on top of the underwater contact. Now that naval areas of interest have shifted from deep water to acoustically more challenging shallower littoral depths, magnetic anomaly detection of submarines is resurfacing as a viable technique. Although manned maritime patrol aircraft and helicopters are still the primary MAD equipped weapons platform, the development of unmanned air vehicles (UAV) and unmanned underwater vehicles (UUV) for MAD surveillance is proceeding at a rapid pace.

The use of unmanned MAD equipped vehicles in large numbers called *swarms* has significant tactical benefits at a relatively low cost. Thirty to forty MAD UAV platforms, operating in a cooperative behavior mode, could potentially patrol 2500 km$^2$ of ocean with a high probability of detecting any submarine in the box and a low probability of false alarm. Each of these vehicles could operate for many hours, and be cheap enough so that they would not have to be recovered, ditching in the ocean when they have exhausted their fuel. To realize these systems, magnetometers that are small and lightweight, inexpensive, low-power, and sensitive are required.

This chapter will present a brief overview of viable magnetic surveillance technologies, and example applications in fixed arrays and mobile submarine detection barriers. Simplified models of geomagnetic and surface wave magnetic noise will be compared to expected target signal strengths for the purpose of ranking the relative importance of noise sources. In addition, a brief description of magnetic noise reduction techniques will be covered.

## 4.2   MAGNETIC HARBOR LOOPS

The first magnetic submarine detection system was the magnetic induction loop or harbor loop. Although there are many configurations and schemes that can be used for induction loop barriers, the design shown in Fig. 4.1 was typical of those deployed during World War II. This geometry is effectively two loops arranged side-by-side with their output voltages subtracted

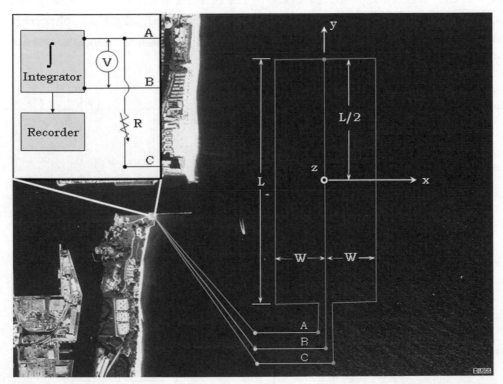

**FIGURE 4.1:** Geometry of an example loop surveillance system designed for the entrance to the harbor at Port Everglades, FL

from each other for the purpose of minimizing the background noise. The resistor placed in the circuit is used to balance the two loops, which comprise the arms in a bridge arrangement. (Some designs use a resistor in both the arms of the bridge circuit for easier operation.) Since the output voltage of the balanced loops is proportional to the rate-of-change in the magnetic field, the signal is integrated before being recorded. The integrator maintains the sensitivity of the system when the magnetic flux linking the loops has a low rate-of-change, typical of slow moving targets.

Due to the present importance of harbor loops in designing detection systems for home-land security and port defense, a more in-depth technical analysis of its operation is warranted. The voltage at the output of the differential loops in Fig. 4.1 will be computed for a magnetized submarine sailing in a straight line across the loops. The voltage $e$ induced in a single loop by a source moving at a velocity $\vec{v}$ can be formulated as [4]

$$e = \oint \left( \vec{v} \times \vec{B} \right) \cdot d\vec{l} \qquad (4.1)$$

where $\vec{B}$ is the magnetic signature produced by the submarine and $d\vec{l}$ is the differential length along the contour of the loop. Representing the source as a triaxial prolate spheroidal dipole whose long axis is in the $x$ direction with moments $(M_x, M_y, M_z)$, the magnetic flux density is given as

$$B_x = \frac{3\mu_0}{4\pi c^3} \left\{ M_x \left[ -0.5 \ln \left( \frac{r+1}{r-1} \right) + \frac{c^2 r}{R_1 R_2} \right] + M_y \left[ \frac{c y \xi}{R_1 R_2 \left( r^2 - 1 \right)} \right] + M_z \left[ \frac{c z \xi}{R_1 R_2 \left( r^2 - 1 \right)} \right] \right\} \qquad (4.2)$$

$$B_y = \frac{3\mu_0}{4\pi c^3} \left\{ M_x \left[ \frac{c y \xi}{R_1 R_2 \left( r^2 - 1 \right)} \right] + M_y \left[ 0.25 \ln \left( \frac{r+1}{r-1} \right) - \frac{r}{2 \left( r^2 - 1 \right)} + \frac{y^2 r}{R_1 R_2 \left( r^2 - 1 \right)^2} \right] \right.$$
$$\left. + M_z \left[ \frac{y z r}{R_1 R_2 \left( r^2 - 1 \right)^2} \right] \right\} \qquad (4.3)$$

$$B_z = \frac{3\mu_0}{4\pi c^3} \left\{ M_x \left[ \frac{c z \xi}{R_1 R_2 \left( r^2 - 1 \right)} \right] + M_y \left[ \frac{y z r}{R_1 R_2 \left( r^2 - 1 \right)^2} \right] \right.$$
$$\left. + M_z \left[ 0.25 \ln \left( \frac{r+1}{r-1} \right) - \frac{r}{2 \left( r^2 - 1 \right)} + \frac{z^2 r}{R_1 R_2 \left( r^2 - 1 \right)^2} \right] \right\} \qquad (4.4)$$

where

$$r = \frac{R_2 + R_1}{2c},$$

$$\xi = \frac{R_2 - R_1}{2c},$$

$$R_1 = \left[(x + c)^2 + y^2 + z^2\right]^{1/2},$$

$$R_2 = \left[(x - c)^2 + y^2 + z^2\right]^{1/2},$$

$$c = \left[a^2 - b^2\right]^{1/2},$$

and $a$ is half of the major axis, while $b$ is half of the minor axis. Mathematical models used to represent the magnetic flux density produced by a submarine can range in complexity from a simple spherical dipole to arrays of dipoles, up to detailed finite element numerical simulations. The prolate spheroidal dipole representation of the target's magnetic field signature is sufficient for explaining the principles behind the induction loop barrier.

In this example, the target is assumed to be sailing at a constant altitude above the loop and with a constant velocity along the $x$ axis. The submarine will be taken to be 90 m long ($2a$), with a beam of 7.5 m ($2b$), and having only a vertical moment ($M_z$) of 150 000 Am$^2$, which is equivalent to the detection example given in [5]. The target will be sailing slowly with a velocity ($v$) of 2 kn (1 m/sec) at altitudes ($z$) of 100 and 200 meters above the loop. The loop length ($L$) is 5 km, and has a half-width ($W$) of 200 m. Placing Eq. (4.2) into (4.1) and numerically integrating for the differential loop configuration in Fig. 4.1, yields the output voltage signal ($V$) shown in Fig. 4.2. As seen in the plot, the peak signal occurs when the target crosses the center conductor. Since the output voltage is directly proportional to the target velocity, signal strengths at other speeds can be computed through direct scaling of these 2 kn results.

The performance of any detection system is ultimately limited by the self-noise level of its sensors. For harbor loops, the two primary sources of self-noise are motion of the loops in the earth's magnetic field and the thermal noise of its conductors. Motion noise can be eliminated by extensive anchoring of the cable or burying it into the sea floor with underwater plows commonly used in the telecommunication industry [6]. The latter will also protect the cables from damage by fishing nets and boat anchors.

The harbor loop's thermal noise, called *Johnson noise*, is produced by the random motion of electrons inside its cables. The root-mean-square (rms) thermal noise voltage, $v_t$, at the output of the loops is given by

$$v_t = \sqrt{4kTR\Delta f} \tag{4.5}$$

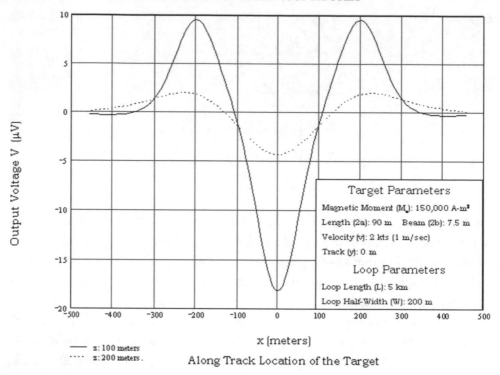

**FIGURE 4.2:** Output voltage signals computed for an example submarine traversing the Port Everglades harbor loop

where $k = 1.3806503 \times 10^{-23}$ J/K (Boltzmann's constant), $T$ is the absolute temperature of the loop in degrees Kelvin, $R$ is the equivalent resistance of the loop in ohms, and $\Delta f$ is the frequency bandwidth in Hz. The equivalent resistance of the loops shown in Fig. 4.1 as measured at its terminals onshore can be computed from

$$R = r\left(\frac{3}{2}(L + l) + W\right) \qquad (4.6)$$

where $r$ is the cable resistance per unit length (ohms/meter), $l$ is the length of the three-conductor cable from the loop connection point to the shore, and L and W have been defined previously. If AWG 10 gauge wire is used in the harbor loop example, $r$ is equal to 3.3 m$\Omega$/meter, $l$ is taken as 6 km, $T$ is equal to 300 K, and using a $\Delta f$ of 0.03 Hz, then the equivalent thermal noise voltage of the harbor loop is computed to be $1.6 \times 10^{-4}$ $\mu$V. The thermal noise voltage is well below the signal voltages, and will not be the limiting factor in detecting the target.

The limiting factor in most submarine detection systems is typically not sensor noise. Instead, exterior noises produced by the natural ocean environment or from manmade sources dominate. Of the naturally occurring noises, geomagnetic and ocean surface wave motion are the primary sources in the electromagnetic frequency band being considered here for submarine detection.

Geomagnetic noise in the 0.0001 to 2 Hz band originates from electric currents in the ionosphere. These currents are produced by decaying magneto-hydrodynamic waves launched by the interaction of solar wind particles with the earth's magnetic field. The magnetosphere's current generates micropulsations in both the magnetic and electric fields at the surface of the ocean. The geomagnetic noise is highly variable and changes with the time of day, season, latitude, and of course with solar activity. The geomagnetic fields are monitored at magnetic observatories located around the world, with daily records available from the U.S. Geological Survey [7].

Average values for the geomagnetic background noise will be used in this example to compute initial estimates of a submarine detection system's noise floor. The average geomagnetic noise power spectral density $G(f)$ will be modeled as [4]

$$G(f) = \frac{f_u f_l}{f^2} \frac{n_g^2}{\Delta f} \tag{4.7}$$

where $n_g$ is the rms geomagnetic fluctuation (a value of 0.3 nT will be used here), $f_u$ is the upper frequency of the passband, and $f_l$ is the lower frequency of the passband. Since an induction loop measures the time rate-of-change in magnetic field over its area, Eq. (4.7) must be multiplied by $(2\pi f)^2$ to get the system's equivalent noise power spectral density of

$$g(f) = (2\pi)^2 f_u f_l \frac{n_g^2}{\Delta f}. \tag{4.8}$$

If the geomagnetic noise is assumed to be uniform over the area of a loop, $A$, then the rms geomagnetic noise voltage $v_g$ measured by the loop can be computed from

$$v_g = 2\pi n_g A (f_u f_l)^{1/2} \tag{4.9}$$

where $A = LW$, and the other terms have been defined previously. For an upper and lower passband frequency of 0.03 Hz and 0.001 Hz, respectively, the expected rms geomagnetic noise voltage at the output of one loop will be about 10 μV.

For the frequency band of this example, even the 200 m water depth is much less than the skin depth in seawater and does not appreciably attenuate the geomagnetic noise. Noise cancellation will therefore be required for reliable operation of the harbor loop barrier, as was the case during World War II. Clearly, geomagnetic noise and its variance is a major consideration during the design phase of any magnetic submarine detection system.

Ideally, when the outputs of the two loops are subtracted, the resultant geomagnetic noise voltage will be zero. However, differences in loop areas and relative orientations, along with near-by sub-bottom conducting anomalies can change the amplitude and phase of the geomagnetic noise seen by the two loops, producing less than perfect cancellation. In practice, some adjustment for the differences in the loops can be made with the balancing resistor shown in Fig. 4.1, but zero noise levels can not be obtained. Passing the output of the loops through the integrator can increase the signal-to-noise ratio by at least a factor of 5, while modern signal processing techniques can achieve even greater improvements.

The second major noise source seen by harbor loops is the magnetic field generated by the ocean's surface waves. As wind driven sea waves and swells move through the earth's magnetic field, small electric currents are induced in the electrically conducting seawater. These currents in turn produce a vertical and horizontal component of a magnetic field perpendicular to the surface wave. However, it will be shown that the wave-induced magnetic field falls off exponentially with distance above or below the sea surface.

The noise voltage at the output of a loop can be determined in a straightforward manner once the magnetic field of the ocean surface wave has been computed. The rms noise voltage $v_w$ for a single loop when the surface wave propagates along the $x$ axis (Fig. 4.1) has been formulated in [4] as

$$v_w = u b_w L_w \left(1 - \cos\left(m W\right)\right)^{1/2} \tag{4.10}$$

where

$$u = \left(\frac{g\lambda}{2\pi}\right)^{1/2},$$

$$m = \frac{2\pi}{\lambda},$$

$$\lambda = \frac{2\pi g}{\omega^2},$$

$$\omega = 2\pi f,$$

and $g$ is the acceleration of gravity (9.8 m/s), $\lambda$ is the wavelength of the surface wave, $L_w$ is the length of the wave's crest (assumed to be $7\lambda$ but not greater than $L$), $f$ is the frequency of the surface wave, and $b_w$ is the rms of the vertical magnetic field generated by the ocean surface wave. (The term $u$ is the velocity of the surface wave and $m$ is its propagation constant.) It is clear from Eq. (4.10) that maximum wave noise is produced when $W$ equals an odd multiple of a half-wavelength, while the minimum occurs at multiples of a full wavelength.

Models of the magnetic field produced by ocean surface waves and swells have been the subject of many investigations. General equations for the horizontal $B_x$ and vertical $B_z$ magnetic fields generated below and above a sinusoidal ocean surface wave were derived in the classic paper [8], and are reformulated here in the *SI* system as

$$B_x = \frac{-A'}{4\pi}\left(\frac{2\left(1+i\beta\right)^{1/2}e^{-md(1+i\beta)^{1/2}}}{1+\left(1+i\beta\right)^{1/2}} - e^{-md}\right) \tag{4.11}$$

$$B_z = \frac{iA'}{4\pi}\left(\frac{2e^{-md(1+i\beta)^{1/2}}}{1+\left(1+i\beta\right)^{1/2}} - e^{-md}\right) \tag{4.12}$$

for fields at a depth $d$ below the water surface, while for fields above the surface at an altitude $h$

$$B_x = \frac{iA'}{4\pi\beta}\left(1 - \left(1+i\beta\right)^{1/2}\right)^2 e^{-mh} \tag{4.13}$$

$$B_z = \frac{-A'}{4\pi\beta}\left(1 - \left(1+i\beta\right)^{1/2}\right)^2 e^{-mh} \tag{4.14}$$

where

$$A' = amF\left(S+iC\right),$$
$$\beta = \frac{\gamma}{m^2},$$
$$\gamma = 4\pi\omega\mu_0\sigma,$$
$$S = \sin\left(I\right),$$
$$C = \cos\left(I\right)\cos\left(\theta\right),$$

and $a$ is the amplitude of the ocean's surface wave (half of the crest-to-trough *wave height*), $F$ is the magnitude of the earth's magnetic field in tesla, $I$ is the angle that the earth's field makes with the horizontal (called the *dip angle*), $\theta$ is the eastward inclination angle from magnetic north for the direction of the surface wave propagation, $\sigma$ is the seawater conductivity in S/m, $\mu_0$ is free space permeability ($\sim 4\pi \times 10^{-7}$ H/m), and all other terms have been defined previously. The rms amplitude of the surface wave's vertical magnetic field at the depth of the harbor loop can be computed from Eq. (4.12) as $b_w = 0.707 B_z$.

The surface wave noise voltage at the output of the harbor loop is a straightforward computation using Eqs. (4.10) and (4.12). Continuing the previous example, and using $\theta = 270°$, $I = 61°$, $F = 51\,000$ nT, $\sigma = 4$ S/m, a high sea state of 6 equal to a wave height of 20 feet ($a = 3$m) and a wave period of 10 sec, produces an rms vertical magnetic field of 0.3 nT at a depth of 100 m. Placing this value into Eq. (4.10) gives a surface wave noise voltage of 3.8 µV for one loop. If the harbor loops were installed in deeper water at a depth of 200 m, the surface wave noise voltage reduces to 0.13 µV.

It is clear that to optimize the design of a harbor loop, the submarine detection system requires detailed attention to many variables. Some of the system parameters will be forced by the ocean bathymetry and environmental conditions at the installation site, and the threat direction to be protected. Other aspects of the design must be selected to maximize the signal-to-noise ratio for expected source strengths of the target and noise levels. The signal-to-noise ratios for target tracks and surface wave propagation directions other than perpendicular to the loops will have to be examined. Although the analysis presented here considered only subtraction of the two side-by-side loops, other configurations and advanced signal processing techniques can significantly increase the system's detection performance.

There are several advantages in using the harbor loop detection system over other barrier concepts. First, the system is very reliable since there are no underwater electronics that can fail and require an at-sea recovery of the component for repair. Harbor loop damage by fishing vessels and recreational boating is highly reduced since the cables would normally be buried in the sea floor. This detection system monitors a continuous boundary leaving no holes in its coverage. To achieve its mission, the target is forced to sail directly over the loop, resulting in a high probability of detection.

There are disadvantages to the harbor loop submarine barrier. First, the installation costs are high since the cable has to be buried in the sea floor or securely anchored to the bottom. Also, the system requires automatic correlation of contacts between the loops and surface surveillance to reduce false-contact reports. Finally, the system has low sensitivity to targets that are outside the loops' perimeter. The harbor loop system is ideal for surveillance of home ports and protected harbors or waterways where it can be permanently installed at a location whose surface traffic is under continuous monitoring.

## 4.3    SUBMARINE BARRIERS USING TRIAXIAL MAGNETIC FIELD VECTOR SENSORS

Several types of triaxial magnetic field sensors are available for consideration in the design of bottom mounted or buoyed arrays. These instruments sense magnetic fields by various physical means, resulting in an assortment of sensitivities, drift, noise, size, weight, power, and cost from which to choose for a barrier's design. Options range from using a few high-performance high-cost transducers with extended detection ranges, or deploying many low-cost short-range sensors to cover the protected waters. Triaxial magnetometers sense the magnetic field vectors in orthogonal directions that can be used to reduce background noises for longer detection ranges, and may also locate a submarine's position in three dimensions.

The application of fluxgate magnetometers in submarine surveillance systems requires higher performance instruments than those used as part of a TDD in a short-range bottom mine. Although the operating principle of a surveillance fluxgate is the same as in a mine, greater care must be taken in the design and fabrication of its magnetic core, windings, and associated electronics. Low drift is much more important while attempting to detect submarines at greater distances, where the period of their signature is much longer.

Modern state-of-the-art fluxgate magnetometers can be specially manufactured to have a very low noise floor, but are more costly. A fluxgate's noise power spectral density $P(f)$ varies approximately as $P(f) = P(1)/f$ nT$^2$/Hz, where $P(1)$ is the noise power at 1 Hz [9]. In general, this equation holds over the range from mHz to kHz frequencies. The DFM24G triaxial fluxgate magnetometer, manufactured as a special order, has a $P(1)$ specification as low as $9 p T^2/Hz$ rms [10].

An array of triaxial fluxgate magnetometers could be used instead of the harbor loops in the previous example. The triaxial magnetic field signatures of the previously described target submarine were computed with Eqs. (4.2) through (4.4) for a single sensor at a target-to-sensor depth of 200 meters over a track length of 2 km centered on the *closest point of approach* (CPA). The rms amplitudes of the three signature components over a 1000 s window around CPA are plotted in Fig. 4.3 as a function of the submarine's track offset from the sensor. (All other

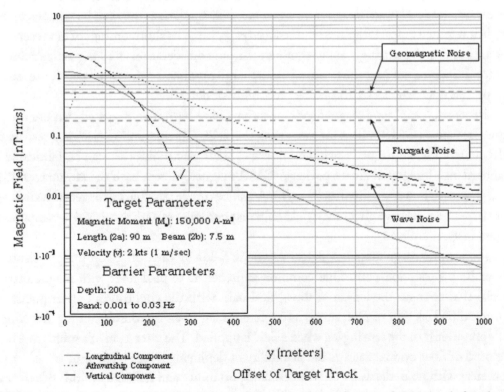

**FIGURE 4.3:** Peak detection field against an example submarine and expected noise levels for a bottom deployed triaxial fluxgate magnetometer

parameters are the same as used previously.) The geomagnetic, surface wave and fluxgate noise levels are also given for comparison. As Fig. 4.3 shows, reduction of the geomagnetic noise is the first priority in extending the detection range of a single sensor node.

The degree of adaptive cancellation of the geomagnetic noise using different sensors in the array depends on the correlation of the noise within the array's environment. Obviously, improved sensor performance will not help until the geomagnetic noise has been reduced. In shallow waters, the ocean surface wave noise may exceed the sensors' noise under some circumstances. A good signal-to-noise ratio is required on all sensor axes if the target is to be localized with a matched filter tracker. Ultimately, the number of array sensors required for protecting an area, the size of the harbor loop example, will depend on the level of noise reduction that can be achieved, and the degree of overlap coverage desired between adjacent instruments.

Besides the fluxgate magnetometer, there are several other types of magnetic vector sensors that might be considered in the design of a submarine barrier. Magnetometers based on the Hall Effect have been used extensively in industrial applications. A Hall sensor detects magnetic fields by passing an electric current through a semiconductor which, in the presence of a magnetic field, produces a voltage in the direction perpendicular to both the current and magnetic field. The advantages of Hall sensors over fluxgates are their small size (they can be manufactured on a single integrated chip), low power and reduced cost, while its shortcomings of lower sensitivity ($\sim30$ nT) and higher drift generally eliminate it from consideration as a long-range submarine detector. However, seeding the ocean with very large numbers of these inexpensive sensors is a concept that may provide an acceptable probability of detection.

Recently, "chip" type magnetic vector sensors have been manufactured that use the giant magnetoresistive (GMR) effect for measuring the field. Changes in the electrical resistance of GMR material with magnetic field are measured with electronics that can be fabricated into a standard size integrated circuit package. GMR magnetometers have the advantages of Hall sensors but with better sensitivities ($\sim1$ nT). Advances made possible by the recently discovered colossal magnetoresistive effect may yield chip magnetometers with sensitivities comparable to fluxgate instruments [11].

Fiber optic magnetometers detect magnetic fields by their effect on the polarization of light or its propagation path. One common approach is to clad fiber with magnetostrictive material that changes the length of the optical path with the application of a magnetic field. An unmodified section of fiber serves as a reference leg in an interferometer for detecting the light's phase shift in the sensing leg when a field is applied. The attraction of a submarine barrier composed of fiber optic sensors is the possibility of deploying a very large array of vector magnetometers with no underwater electronics. The portability and reliability of this system would be significant. Unfortunately, the sensing leg of the present fiber magnetometers is also sensitive to temperature, pressure, acoustic, and hydrodynamic influences of the ocean background that generally swamps the magnetic signal of interest.

The most sensitive magnetometer, which also happens to be a vector sensor, is the superconducting quantum interference device (SQUID). This instrument uses the Josephson Effect to measure very small changes in the magnetic field. There are many different SQUID designs based on low or high temperature superconducting material. However, in all cases the superconducting portion of the sensor must be kept at cryogenic temperatures. Although DC sensitivities much less than 1 pT are typical for a SQUID magnetometer, their cryogenic requirements make it difficult to deploy them in a submarine barrier for an extended period of time.

The standard method of reducing geomagnetic background noise is through direct cancellation with a reference station. An extra sensor is placed at a distance from the detection array and is used as the noise reference in standard cancellation algorithms. The noise reference should be located far enough from the barrier array so that little of the target's signature is seen by the reference and inadvertently subtracted, but it must be close enough to ensure good noise correlation with all sensors in the array. If the detection array is sufficiently separated spatially, then the need for an additional noise reference may be eliminated.

Least-squares minimization is the most direct method for canceling geomagnetic background noise. Let one axis of a detection sensor be denoted by $H_d$, and the three axis of the noise reference be given by $N_x$, $N_y$, and $N_z$. A noise minimization function can be written as

$$\varepsilon\left(a, b, c\right) = \left(H_d - a\,N_x - b\,N_y - c\,N_z\right)^2 \qquad (4.15)$$

where $a$, $b$, and $c$ are coefficients to be determined that minimizes $\varepsilon$ for all measurements taken during array calibration for noise reduction. (The process is repeated for each axis of each sensor in the array.) Once the three coefficients have been computed, they are fixed and used during the array's surveillance mode by continually comparing the magnitude of $\varepsilon$ against a threshold. This technique can be used in the time domain with band limited data, or in the frequency domain where the three coefficients are computed for each bin in the spectrum. Although the ultimate degree of noise cancellation depends on the geomagnetic noise correlation between the sensors, noise reductions greater than 20 db can be easily achieved.

The primary advantage of a triaxial magnetic submarine barrier over the harbor loop is the ability of a multi-axis sensor system to accurately track a target in three dimensions. In addition, the small volume of the fluxgate sensor and the reduced size of its interconnecting cables between *sensor nodes* make this array much more portable than a harbor loop. The drawback of a bottom deployed fluxgate array is its susceptibility to damage and the failure of its underwater electronics, which may force recovery of the units increasing the system down time and costs to repair.

## 4.4   SUBMARINE BARRIERS USING TOTAL FIELD MAGNETOMETERS

For the fixed bottom array, it was assumed that the triaxial vector magnetometers are mechanically stable and do not move. This assumption must hold to a very high degree. If, in the previous example, one of the magnetometer's axis is normal to the earth's field (worse case), then the noise $n_r$ produced by its rotation through an angle $\theta_r$ is given by $n_r = F \sin(\theta_r)$, which for small angles becomes $n_r = F\theta_r$. When $F$ is 51 000 nT and $\theta_r$ is as small as $0.001°$ then $n_r$ is 0.9 nT, well above any other noise sources.

Motion noise is especially high when a vector magnetometer is buoyed from the surface, suspended in the water column, or attached to a moving platform. If a vector sensor's motion noise is in the passband of interest, then it can not be eliminated through filtering. For these applications, the triaxial magnetometer must generally be used in a *total field* configuration by computing the vector sum of its components. However, some signal information is lost with a total field sensor owing to the presence of the much larger earth's magnetic field. This is an important point that requires further explanation.

The derivation of the total field component of a small signal in the presence of the much larger earth's magnetic field begins with the vector sum of their components. The vector sum of the three signal components $(s_x, s_y, s_z)$ in the presence of the earth's field $(F_x, F_y, F_z)$ is given by

$$T = \sqrt{\left(F_x + s_x\right)^2 + \left(F_y + s_y\right)^2 + \left(F_z + s_z\right)^2} \qquad (4.16)$$

which can be expanded to

$$T = \sqrt{F_x^2 + s_x^2 + 2F_x s_x + F_y^2 + s_y^2 + 2F_y s_y + F_z^2 + s_z^2 + 2F_z s_z}. \qquad (4.17)$$

Combining terms

$$T = \sqrt{|\vec{F}|^2 + 2F_x s_x + 2F_y s_y + 2F_z s_z + |\vec{s}|^2} \qquad (4.18)$$

where $|\vec{F}|$ and $|\vec{s}|$ are the magnitudes of the earth's field and the signal, respectively. Factoring $|\vec{F}|^2$ from inside the radical, and expanding the result as a series gives

$$T = |\vec{F}| \left(1 + \frac{F_x s_x}{|\vec{F}|^2} + \frac{F_y s_y}{|\vec{F}|^2} + \frac{F_z s_z}{|\vec{F}|^2} + \frac{|\vec{s}|^2}{2|\vec{F}|^2} + \cdots \right). \qquad (4.19)$$

Multiplying through by $|\vec{F}|$ and, under the small signal assumption, neglecting all terms on the order of $|\vec{s}|/|\vec{F}|$ and higher, Eq. (4.19) becomes

$$T \cong |\vec{F}| + e_x s_x + e_y s_y + e_z s_z \qquad (4.20)$$

where $e_x$, $e_y$, $e_z$ are unit vector components in the direction of the earth's field. Subtracting the magnitude of the earth's field from Eq. (4.20) and writing the result in its general vector form gives

$$T \cong \hat{e} \cdot \vec{s}. \qquad (4.21)$$

This equation shows that a total field sensor, regardless of its sensing mechanism, measures magnetic field signatures primarily in the direction of the much larger earth's field, and under these circumstances can not resolve components normal to it.

The above analysis assumes that the three components of the magnetometer are perfectly orthogonal, have the same gain, and have no DC offsets. This of course is never the case. Even using the high performance DFM24G triaxial fluxgate magnetometer, total field (vector sum) noise levels larger than several hundred nT are observed when the sensor is allowed to freely rotate in the earth's field. Corrections for the sensor inaccuracies can be made mathematically with a special calibration process.

Techniques to correct a triaxial magnetometer's orthogonality, gain, and offset for use as a total field sensor have been described by several investigators, the most recent of which can be found in [12] and [13]. Here, a simple example will be given to demonstrate order-of-magnitude reduction in motion noise that might be realized. The corrected total magnetic field vector sum, $T_c$, is given by

$$T_c = \sqrt{h(H_x - O_x)^2 + (a(H_x - O_x) + b(H_y - O_y))^2 + (c(H_x - O_x) + d(H_y - O_y) + e(H_z - O_z))^2}$$

$$(4.22)$$

where $H_x, H_y, H_z$ are the triaxial fields measured by the uncorrected magnetometer, and $a$, $b$, $c$, $d$, $e$, and $h$ are corrections for both orthogonality and gain errors combined, while $O_x, O_y, O_z$ are coefficients for correcting the offset errors. The 9 correction coefficients in Eq. (4.22) are determined experimentally as a part of the sensor calibration process.

To calibrate a total field vector magnetometer, its triaxial field readings are recorded simultaneously as the sensor is rotated about each of its axes. The calibration can be performed either inside a large Helmholtz coil with a known applied magnetic field, or the instrument can be rotated within the earth's field using a nearby reference sensor to monitor the changes in the background. Rotations through $360°$ should be used for each axis to minimize experimental errors and to avoid singular solutions. After all the data are collected, the correction coefficients can be found by minimizing an error function such as $\varepsilon = \max |T_t - T_c|$, where $T_t$ is the true total field applied during each measurement of $H_x, H_y, H_z$ as the sensor is rotated.

Many mathematical software packages contain routines that can iteratively solve for the 9 correction coefficients that minimizes $\varepsilon$ over all calibration angles. However, care must be taken to find the correction coefficients that globally minimizes $\varepsilon$ for best performance. A mathematical simulation of a triaxial magnetometer rotated within a constant background field can be used to show what might be achieved with orthogonality, gain, and offset corrections.

In this example, the triaxial magnetometer's errors will be set at $\pm 0.1^0$ for orthogonality, $\pm 5$nT for offsets, and linear gain errors of $\pm 10\%$. To make the simulation more realistic, a uniform random sensor noise level of $\pm 0.1$nT will be added to each measurement axis. The sensor was calibrated by mathematically rotating it in three directions through 1080 different angles within a 55 000 nT background field. The 9 correction coefficients in Eq. (4.22) were found with a non-linear iterative routine that minimized $\varepsilon$. These correction coefficients were then used to compute $\varepsilon$ while rotating the sensor through other 100 000 angles for testing the orthogonality correction. The maximum error in the total field was found to be more than 500 nT before the axis correction, which was reduced to less than 0.25 nT after calibration.

Magnetic field sensing mechanisms exist that can measure the absolute total field directly. These sensors are sometimes called *scalar magnetometers* since they can only measure the magnitude of the total magnetic field and not its direction. A scalar magnetometer always measures a small signal's component in the direction of the larger earth's magnetic field according to Eq. (4.21), regardless of frequency.

Nearly all scalar magnetometers sense magnetic fields using a phenomenon called *nuclear magnetic resonance*. These sensors excite protons or electrons with a DC or AC magnetic field, and then measure their precession frequency as they gyrate about the external field to be measured. The *gyromagnetic ratio* of the precession frequency to the applied magnetic field is a constant, which for a proton is $2.67515255 \times 10^8$ radians/(s-T). The high sensitivity of the magnetic resonant magnetometer results from precise frequency measurements made possible with modern electronic counters. The proton's gyromagnetic ratio is a constant used in the SI system of units to define a measure of magnetic field (tesla) and electric current (ampere). Although not required for submarine detection, a properly designed scalar magnetometer can measure a magnetic field with a high absolute accuracy.

The three general types of scalar resonant magnetometers are the proton precession, Overhauser, and the optically pumped. A detailed explanation of their operating principles and design considerations can be found in [14]. The advantages and disadvantages of their application to submarine detection may be understood with a simple description of their operation.

Within the proton precession magnetometer, a sample of proton-rich liquid (such as kerosene) is subjected to a strong DC magnetic field that aligns some of them in the direction of the applied field. When the strong bias field is removed, the protons precess about the external field, which in this case is the vector sum of the large earth's background field and the much smaller submarine's signature when present. Since the proton has a magnetic moment and a field of its own, its precession frequency can be measured with an induction pickup coil and frequency counter. The amplitude of the magnetic field is resolved by relating the measured frequency to the gyromagnetic constant. The requirement for a large polarization bias field makes the simple proton precession magnetometer a higher power device.

The Overhauser Effect has been used to reduce the power requirements of the proton precession sensor. This effect involves the transfer of the magnetic polarization of a material's electrons to its protons. Since electrons are much easier to polarize than protons, Overhauser magnetometers require less energy than the simple proton precession sensors.

Optically pumped scalar magnetometers are more sensitive, require less power, and have a higher bandwidth than proton precession instruments. These sensors exploit the properties of electron spin resonance directly to quantify magnetic fields. Electrons in the atoms of $He^4$ or vaporized alkali metals ($Na^{23}$, $K^{39}$, $Rb^{85}$ & $Rb^{87}$, and $Cs^{133}$) are optically pumped into higher energy states, from which they decay. Since the spin rate of electrons is 600 times that of protons, optically pumped scalar magnetometers can be constructed that are more sensitive and have a better frequency response than other spin resonant sensors.

High sensitivity and reduced motion noise makes the nuclear magnetic resonance and optically pumped scalar magnetometers attractive for submarine detection from moving platforms. Naturally, there are advantages and disadvantages to each type of total field sensor. Manufacturers are continually developing innovative approaches to improve the sensitivity of their instruments, while at the same time reducing their unfavorable characteristics. Besides sensitivity and cost, total field sensor specifications that should be considered in the design of a submarine surveillance system include power, weight, volume, bandwidth, gyroscopic motion noise, susceptibility to high field gradients, orientation requirements relative to the earth's field, calibration and maintenance, and ease in operation. The relative importance of these parameters will depend on the deployment platform's size, available power, self-noise, maneuverability, vibration, degree of automation, mission time, and reusability requirements.

Noise levels as low as 0.2 pT has been reported for $He^4$ scalar magnetometers [15]. If it were possible to reduce all magnetic noise down to a 0.2 pT sensor noise level, the maximum detection range for the submarine target source strength used in the previous examples would be on the order of 4 km. Achieving system performance at this level, however, is a major challenge.

Manned maritime patrol aircraft generally use optically pumped $He^4$ or $Cs^{133}$ magnetometers in their MAD systems. The sensor is housed in either a boom that extends behind or in front of a fixed wing aircraft or is housed in a pod that is pulled several meters behind a helicopter. The purpose of the pod and extension boom is to remove, as much as possible, the sensor from the magnetic noise generated by the aircraft itself. Unfortunately, the aircraft noise sensed by the remote sensor is generally still too high and requires additional reduction.

Unlike in-water submarine barriers, geomagnetic and surface wave noise is not as dominant for airborne surveillance systems. Because the aircraft is traveling at high speed in comparison to the target, the energy of the detected signal resides in a band above the stronger geomagnetic noise that plagues bottom mounted arrays. Also, the aircraft can fly at an altitude where the surface wave noise is reduced considerably. However, natural magnetic anomalies created by geologic sources within the earth can mask targets or produce false contacts. Geologic noise is typically addressed by precursor airborne magnetic surveys and mapping areas of interest. Magnetic contacts that correlate with mapped anomalies can then be discarded as false. A more complete description of noise sources seen by MAD detection systems can be found in [15].

In spite of mounting magnetic sensors on extended booms and inside trailing pods, the limiting noise sources for airborne MAD systems originate from the aircraft's structure or onboard electric systems. As was discussed in Chapter 2 for ships and submarines, a sensor platform's permanent and induced magnetization, eddy currents generated in conducting material, and the stray fields caused by powered electric circuits can generate magnetic fields at the sensor that can swamp the target's signal. However, the platform's magnetic field does not have to be reduced in the entire volume around it, but only at the location of the sensor.

As is the case for naval vessels, the first step in the reduction of the magnetic fields on any sensor platform is to eliminate, as much as possible, the sources before considering active cancellation techniques. The permanent and induced magnetization can be minimized by constructing the vehicle and its component's with non-ferrous material. If the airframe and internal structure are also non-conducting, then eddy current generated fields will be curtailed. Care in the design of onboard electric powered systems (cables, motors, generators, batteries, etc.) can significantly reduce stray fields. Monitoring of currents in important circuits can be used to cancel any remaining stray field components. Typically, active cancellation of any leftover induced, permanent, and eddy current noise is accomplished either numerically or with special purpose hardware.

The technique used to cancel the aircraft noise at the location of its scalar magnetometer is based on an algorithm developed originally by Tolles and Lawson [16]. The total magnetic noise produced by the sensor platform, $N_a$, can be expressed as

$$N_a = N_p + N_i + N_e \qquad (4.23)$$

where $N_p$ is the noise from its permanent magnetization at the location of the detection sensor, $N_i$ is the noise from induced magnetization, and $N_e$ is the eddy current noise generated by the platform during rotational maneuvers. If these three platform noises can be estimated in real-time and subtracted from the detection sensor's readings, then the performance of the MAD system could be improved.

Although the direction and magnitude of the platform's permanent magnetization can be assumed constant over the duration of the mission, the magnetic noise signal detected by the scalar magnetometer contains only those field components aligned in the direction of the earth's magnetic field according to Eq. (4.21). Therefore, the platform's permanent magnetization noise signal ends up being a function of its orientation within the earth's field even though the source itself is constant. This noise can be represented as

$$N_p = \sum_{i=1}^{3} p_i u_i \qquad (4.24)$$

where $p_i$ is the noise of vector field components produced by the platform's permanent magnetization in its transverse ($i = 1$), longitudinal ($i = 2$), and vertical directions ($i = 3$), and $u_i$ are the direction cosines between the platforms axis and the direction of the earth's field. The direction cosines are computed from $u_i = \cos(\alpha_i)$, with $\alpha_i$ being the direction angles between the earth's field and the platform's three axes.

For the induced magnetic noise, both the magnetization itself and its resultant noise field as seen by the scalar magnetometer are functions of the platform's direction angles. The induced noise is given by

$$N_i = \sum_{i=1}^{3} \sum_{j=1}^{3} a_{ij} u_i u_j \qquad (4.25)$$

where $a_{ij}$ are constants to be determined empirically. Since $u_i u_j = u_j u_i$, then $a_{ij} = a_{ji}$, and the number of independent constants is reduced to six.

In general, it is not necessary to solve for the earth's constant field, which is removed from the measurements with a high-pass filter. Under this condition, and since

$$\sum_{i=1}^{3} u_i^2 = 1, \qquad (4.26)$$

$u_3^2$ can be eliminated, allowing $a_{33}$ to be set to zero. The number of independent unknown constants for the induced noise is reduced to five.

The eddy current sources induced in the platform as it rotates in the earth's magnetic field are proportional to the time derivative of the direction cosines. These sources in turn generate a magnetic noise field that must be projected into the direction of the earth's field. Therefore, the eddy current noise field can be written as

$$N_e = \sum_{i=1}^{3} \sum_{j=1}^{3} b_{ij} u_i \dot{u}_j \qquad (4.27)$$

where $\dot{u}_j$ is the time derivative of $u_j$, and $b_{ij}$ are constants to be determined empirically. Taking the time derivative of Eq. (4.26), $u_3 \dot{u}_3$ can be eliminated in Eq. (4.27) allowing $b_{33}$ to be set to zero. This leaves eight independent unknown constants for the eddy current noise.

The sixteen unknown noise constants are determined empirically at the beginning of a mission. The aircraft is maneuvered through angular rotations along several different headings relative to the direction of the earth's magnetic field. These maneuvers are generally conducted at higher altitudes to minimize contamination from ocean surface wave noise and geologic noise sources. Typically, the direction cosines are computed from magnetic fields measured by a triaxial fluxgate vector sensor. The magnetic noise seen by the aircraft's scalar magnetometer is then correlated to the simultaneously measured direction cosines and their rate-of-change. After the sixteen noise parameters are computed from the *MAD compensation measurements* (MADCOMP), the coefficients are used in real-time to cancel the aircraft noise as it searches for submarines at a lower altitude. Although the Tolles-Lawson noise compensation algorithm has traditionally been applied to manned MAD aircraft, in principle, it can be employed on mobile unmanned surveillance platforms as well.

# REFERENCES

[1]   D. Kennedy. (2003. Dec.). The midget submarine attack against Sydney: May 1942. Mysteries/Untold Sagas of the Imperial Japanese Navy. [Online]. Available: http://www.combinedfleet.com/Tully/sydney42.html.

[2]   R. Walding. (2003. Oct.). Indicator loops around the world. Moreton Bay College. Queensland, Australia. [Online]. Available: http://home.iprimus.com.au/waldingr/loops.htm.

[3]   R. Burgess, "Lest we forget," *Proc. U.S. Naval Inst.*, vol. 128/7/1,193, July 2002.

[4]   D. G. Poivani, "Magnetic loops as submarine detectors," Poster L-1, MTS-IEEE Oceans '77.

[5]   G. K. Hartmann and S. C. Truver, *Weapons That Wait*, 2nd ed, Annapolis, MD: Naval Institute Press, 1991, p. 115.

[6]   J. Chesnoy, *Undersea Fiber Communication Systems*. London, United Kingdom: Academic Press, 2002, pp. 514–515.

[7]   J. E. Caldwell (2005. May). Real-time geomagnetic data. USGS. Reston, VA. [Online]. Available: http://geomag.usgs.gov/

[8]   J. T. Weaver, "Magnetic variations associated with ocean waves and swell," *Jour. Geo. Res.*, vol. 70, no. 8, Apr. 1965.

[9]   P. Ripka, *Magnetic Sensors and Magnetometers*, Boston, MA: Artech House, 2001, p. 105–109.

[10]  B. Billngsley (2005). DFM24G 28 bit resolution serial digital triaxial fluxgate magnetometer. Billingsley Aerospace & Defense. Germantown, MD. [Online]. Available: http://www.magnetometer.com/products/specs/dfm28g.pdf.

[11]   P. Ripka, *Magnetic Sensors and Magnetometers*, Boston, MA: Artech House, 2001, p. 129–169.

[12]   S. Takagi, J. Kojima, and K. Asakawa, "DC cable sensors for locating underwater telecommunication cable," *Proc. MTS-IEEE Oceans '96*, vol. 1, 23–26 Sept. 1996, pp. 339–344.

[13]   R. E. Bracken, D. V. Smith, and P. J. Brown, "Calibrating a tensor magnetic gradiometer using spin data," USGS. Reston, VA. [Online]. Available: http://www.usgs.gov/pubprod/.

[14]   P. Ripka, *Magnetic Sensors and Magnetometers*, Boston, MA: Artech House, 2001, p. 267–304.

[15]   L. Bobb, J. Davis, G. Kuhlman, R. Slocum, and S. Swyers, "Advanced sensors for airborne magnetic measurements", *Proc. 3rd International Conference on Marine Electromagnetics (MARELEC)*, July 2001.

[16]   S. H. Bickel, Small signal compensation of magnetic fields resulting from aircraft maneuvers, *IEEE Trans. Aero. and Elect.*, vol. AES-15, no. 4, Jul. 1979.

CHAPTER 5

# Summary

The magnetic field signatures of naval vessels have been exploited by weapons and detection systems for over 80 years. The primary source of a ship's magnetic field is the ferromagnetic steel used in the construction of its hull, internal structure, and onboard machinery and equipment. The initial application of iron cladding over the wooden hulls of combatants, and later their entire construction from steel, originated as a countermeasure to damage from naval artillery. However, this countermeasure produced magnetic fields that resulted in the development of a new weapon—the magnetic influence mine. Simultaneously, the magnetic field signatures of submarines were exploited by underwater and airborne surveillance systems for their detection and localization.

Future naval weapons may further exploit a ship's magnetic field for terminal homing and guidance, and for proximity fusing. Since the magnetic field falls off relatively fast with distance, its detection by a weapon system's fuse ensures that the warhead is in the near proximity of the target when it detonates. This application will be enhanced by the miniaturization of magnetic field sensors, which is progressing at a rapid pace.

Iron is the primary ferromagnetic alloy in naval steels. Ferromagnetic elements have unpaired electrons in their 3d orbits, and must also be spaced in their crystalline structure at an optimum distance for a favorable exchange of energy. Non-magnetic austenitic stainless steel with high chromium content has an atomic spacing that is not favorable for ferromagnetism, but still keeps its desirable properties as a protective armor. Therefore, non-magnetic steel is an attractive replacement for high carbon magnetic alloys, and could significantly reduce the magnetic signatures of ships.

The earth's natural magnetic field induces a magnetization in a ship depending on its latitude, longitude, and heading. The induced magnetization can be broken into three orthogonal components that are parallel to the vessel's vertical, longitudinal, and athwartship axis. Each of the three induced magnetizations in turn generates their own characteristic flux distributions around the hull. Mechanical stress on the ship's ferromagnetic structure will cause some of the induced magnetization to be retained as a permanent component, which does not immediately change with the earth's inducing field. Similarly, the permanent magnetization can be separated into the ship's three orthogonal directions and also produce their own characteristic signatures. The total number of ferromagnetic signature components comes to 18; three signature components for each of the three directions of induced and permanent magnetization.

In addition to ferromagnetism, other important shipboard sources of magnetic fields are eddy currents, corrosion related fields, and stray fields. Eddy currents and their associated magnetic fields are generated primarily in the hull of a vessel as it rolls in the earth's magnetic field. Corrosion related fields originate from natural corrosion currents that flow between electrochemically dissimilar metals along a ship's hull, or between cathodic protection anodes and the materials they are designed to protect from rusting. The shipboard sources of stray fields are electric machinery and power distribution systems that carry electric current in circuits that form loops. These secondary magnetic field sources have both DC and AC components.

A large portion of the magnetic field signatures of surface ships and submarines can be eliminated by constructing them from non-magnetic and non-conducting materials. Such a naval vessel would have little or no ferromagnetic materials to distort the earth's background field, no eddy current fields since the ship's structure would be non-conducting, and little corrosion currents fields owing to the non-conducting properties of its construction materials. However, reducing or eliminating the onboard sources of stray fields will require that high current electric machinery and power distribution systems be designed up-front with a low magnetic field requirement. Since it is unlikely that all shipboard sources of the magnetic field will be completely eliminated, active compensation of the remaining components can provide additional reduction in signature and the vessel's resultant susceptibility to mines and detection systems.

Bottom magnetic influence mines were developed to counter mechanical minesweeping systems that proved to be effective in clearing moored mines during World War I. Magnetic sensors used in influence mines deployed during World War II were based on either dip needle principles or magnetic induction in a solenoid. Later, low power fluxgate magnetometers were employed as target detection devices.

Modern mines have improved capabilities in several areas. Some of the important advancements are:

1.  increased lethality and damage radius provided by more powerful explosives;
2.  self propelled versions with larger attack ranges that increase the threat level of a minefield while requiring fewer mines to be deployed;
3.  better classification of target types and ship classes; and
4.  increased resistance to countermeasures such as mine hunting, mine sweeping, and target ship signature reduction.

The last two of these improvements are made possible by developments in sensor technologies. There are four main objectives of the influence sea mine:

1.  Eliminate or significantly reduce ambient, natural, or manmade background noise.

2.  Detect the presence of a surface ship or submarine within its attack range and measure its signatures.

3.  Recognize mine countermeasure signals as false targets and inhibit from firing.

4.  Identify those signatures coming from valid targets and to detonate its explosive warhead at a time that meets its lethality requirement.

These objectives are met on a statistical basis by using outputs from one or more sensors that detect hydrodynamic pressure changes, acoustic/seismic signals, and magnetic fields and their gradients. Additional improvements can be incorporated into the weapons with signal processing techniques made possible by modern microprocessors with low power demands.

To damage a target ship, an influence mine is not required to make contact when it explodes. The formation of a high-pressure gas bubble by the charge's detonation forms a shock wave and a pressure pulse that impinges on the ship's hull. If the hull can not completely absorb the shock energy through plastic deformation, it will rupture. Even if a vessel's hull does not split, the rapid acceleration of the ship's structure from a shock wave can injure the crew, severely damage the machinery and propulsion systems, weapons, and electronic devices.

The optimum sensitivity threshold setting at which a mine detonates must be selected to impart the required degree of damage to all primary targets that pass within its lethal range. At the same time, the actuation setting can not be too sensitive, otherwise the mine will detonate on targets beyond its damage range and be wasted, or the mine can be swept easily and decoyed into firing with no target present. The optimum mine sensitivity setting is one that will produce an actuation distance that is no greater than the damage range of its explosive charge.

In addition to mine sweeping, hunting and ship signature reduction form two of the three defensive layers against influence mines. The first and best defense against mines is to prevent their manufacture, transport, and deployment. If somehow mines are deployed in contested waters, their detection through hunting, destruction with explosive charges, and decoying them into harmless detonation with influence sweeping comprise the second defensive layer. However, if one or two weapons remain inside the transit areas at the completion of mine-clearing operations, then ship protection must rely on underwater signature control as the third layer of defense.

There is a direct relationship between mine-sweeping effectiveness and ship signature reduction. The sensitivity setting of a mine is selected to match its maximum actuation distance to the mission abort damage range of its explosive charge for the expected priority target. If a minefield planner is unaware that the target vessel has a reduced signature, or if he ignores this fact and does not increase his mines' sensitivity, then the target ship might sail through the minefield without actuating any weapons. This is called a catastrophic failure of the minefield. If, however, the minefield planner does increase the mines' sensitivity to re-match their actuation and damage ranges, then the minefield is easier to sweep, resulting in larger swept widths and requiring fewer passes (less time and/or fewer minesweepers) to clear the field.

Future underwater stealth technologies may be able to reduce a vessel's magnetic field signature low enough to completely hide it from a mine, even at shallow water depths. In addition, it may be possible to blind a mine with a jamming signal that prevents it from coming to a correct firing decision while a valid target sails past it. When the underwater signatures of naval vessels are reduced, unsophisticated mines are more easily swept or become ineffective, while modern sensitive and sweep-resistant mines are easier to jam. For the first time since the invention of the magnetic naval mine, countermeasures may be inching out in front of mine development, even if only by a small margin.

Reducing the magnetic field signatures of submarines not only diminishes their susceptibility to influence mines, but it also decreases their detectability from underwater and airborne surveillance systems. During World War II, large inductive loops were placed inside and at the entrance to important harbors as an alert against enemy submarines that might attempt to slip in and sink anchored ships. In addition, patrol aircrafts were equipped with magnetic sensors to detect submerged submarines sailing through choke points such as the Straits of Gibraltar. Both of these surveillance schemes were successful in detecting submarines that resulted in their sinking.

Although the magnetic induction loop barrier has its origin in World War II, it is still applicable against modern threats as a component in homeland security, force and harbor protection, drug interdiction, and in the monitoring of coastlines. Loops more than 10 km in length were deployed and operated successfully during World War II. Harbor loops of much greater length may be possible today with modern technology. The advantage of this submarine detection system is its ability to monitor a large shallow water area on a permanent basis with little, if any, underwater maintenance required due to the absence of underwater electronics.

One disadvantage of the harbor loop surveillance system is its inability to localize a target outside its physical extent. Triaxial magnetic vector sensors, such as fluxgate magnetometers, provide much more information that can be used in tracking a target in three dimensions. However, it may be expensive or difficult to repair or replace the underwater electronics of this type of sensor array when there is a failure. Other vector instruments such as the Hall probe, GMR, fiber optic, and SQUID magnetometer all have issues that make them unattractive as bottom mounted sensors in a submarine surveillance system.

When used as a total field sensor, triaxial magnetometers must be corrected for orthogonality, gain, and offset errors before computing the vector sum of their components. Total field magnetometers are used in buoyed arrays or on moving platforms to reduce the noise caused by the rotations of a vector magnetic sensor within the earth's field. Magnetic sensors based on the nuclear magnetic resonance phenomenon measure the total field directly. However, one drawback of the total field sensor is that it primarily detects the component of the submarine's small signature that is in the direction of the much larger earth's magnetic field. Small signatures that are normal to the earth's field are virtually invisible to the total field sensor.

There are several sources of magnetic noise that must be addressed in an underwater or airborne submarine surveillance system. Geomagnetic noise caused by solar winds that impinge on the earth's ionosphere generates electric currents and associated magnetic field variations. Also, movement of conducting seawater through the earth's magnetic field by surface waves and swells generates electric currents and magnetic noise. However, both of these noise types can be reduced with a background noise reference sensor that subtracts or adaptively cancels the noise in the surveillance magnetometers.

When magnetic field sensors are mounted on moving platforms, such as an aircraft, additional noise sources appear. Geologic noise and the induced, permanent, eddy-current, and stray fields of the moving sensor platform itself are noise sources that are not generally issues with fixed bottom mounted arrays. At present, geologic noise is removed by premapping the areas of interest and charting large natural magnetic anomalies before surveillance operations begin. The self-noise of the sensor platform is reduced by constructing it, as much as possible, from non-magnetic and non-conducting materials. Any residual ferromagnetic or eddy-current generated noise is removed from the surveillance sensor using an adaptive noise cancellation technique called the Tolles-Lawson algorithm. Stray field noise is reduced through proper up-front design of the platforms electrical systems and adaptive noise cancellation using a current monitor in the onboard circuits as a noise reference.

Magnetic detection systems have been proposed for unmanned air, surface, and underwater vehicles that would search for submarines or other underwater magnetic targets. Depending on the mission, these surveillance platforms could operate alone or in large numbers called *swarms*, and would improve the probability of detecting an acoustically quiet target in a high noise shallow water environment. In general, it should be easier and less expensive to remove sources of magnetic noise from a small unmanned sensor platform than for large manned systems, reducing the performance requirements of the Tolles-Lawson compensation algorithm or eliminating it altogether.

# Author Biography

Dr. John J. Holmes received his B.S. (1973), M.S. (1974), and Ph.D. (1977) degrees in electrical engineering from West Virginia University. He joined the Naval Surface Warfare Center (1977) and is currently the Senior Scientist for the Underwater Electromagnetic Signatures and Technology Division, where he is responsible for the development of underwater electromagnetic field signature reduction systems for surface ships and submarines. Dr. Holmes has written 24 peer-reviewed papers, holds 10 patents, received the David Packard Excellence in Acquisition Award (1999), the Meritorious Civilian Service Award (1986), and is a senior member of the Institute of Electrical and Electronic Engineers.

Printed in the United States
by Baker & Taylor Publisher Services